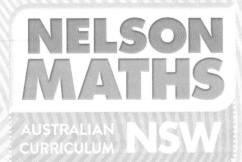

Student Book

Glenda Bradley

NELSON
CENGAGE Learning

Australia • Brazil • Japan • Korea • Mexico • Singapore • Spain • United Kingdom • United States

Contents

Counting by 5s

You will need: coloured pencils

1 a Colour the first box yellow. Count on 5 and colour that box yellow. Continue to count on by 5s and colour each box.

b What final digit pattern do you see? _____

2 a Colour the third box green. Count on 5 and colour that box green. Continue to count on by 5s and colour each box.

b What final digit pattern do you see? _____

0	1	2	3	4	5	6	7	8	9
10	11	12	13	14	15	16	17	18	19
20	21	22	23	24	25	26	27	28	29
30	31	32	33	34	35	36	37	38	39
40	41	42	43	44	45	46	47	48	49
50	51	52	53	54	55	56	57	58	59
60	61	62	63	64	65	66	67	68	69
70	71	72	73	74	75	76	77	78	79
80	81	82	83	84	85	86	87	88	89
90	91	92	93	94	95	96	97	98	99

3 Finish the patterns.

a 24, 29, 34, 39, _____, _____, _____, _____, _____, _____, _____

b 56, 61, 66, 71, _____, _____, _____, _____, _____, _____, _____

Counting by 3s

1 Begin at 1 and continue to write the numbers in the table until you reach 24.

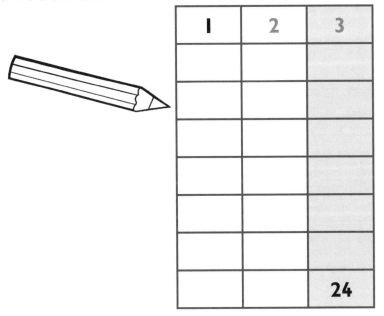

1	2	3
		24

2 a Record the numbers from the last column.

_____

b What are they counting by? _____

c Write the next 4 numbers in the counting pattern.

_____

3 a Record the numbers from the first column.

_____

b What are they counting by? _____

c Write the next 4 numbers in the counting pattern.

_____

d How did you work out the numbers?

_____

Unit 1   **Counting** (TRB pp. 22–25)
Whole numbers MA1-4NA applies place value, informally, to count, order, read and represent two- and three-digit numbers

5

Counting Patterns

You will need: a calculator

1 Complete the number patterns.

a 1, 6, 11, 16, 21, _____, _____, _____, _____, _____

b 3, 13, 23, 33, 43, _____, _____, _____, _____, _____

c 86, 84, 82, 80, 78, _____, _____, _____, _____, _____

d 95, 85, 75, 65, 55, _____, _____, _____, _____, _____

e 9, 11, 13, 15, 17, 19, _____, _____, _____, _____, _____

f 3, 6, 9, 12, 15, _____, _____, _____, _____, _____

g 98, 88, 78, 68, 58, _____, _____, _____, _____, _____

h 78, 76, 74, 72, 70, _____, _____, _____, _____, _____

2 Use a calculator to help you work out the counting pattern.

- Press $\boxed{5}$ $\boxed{4}$. Now press $\boxed{-}$ $\boxed{3}$ and then $\boxed{=}$.

- Write the number in the first space below.

- Keep pressing $\boxed{=}$ to find the other numbers in the pattern, and write them in the spaces.

54, _____, _____, _____, _____, _____, _____,

_____, _____, _____, _____

3 Use the calculator to make your own pattern. Record it below.

_____, _____, _____, _____, _____, _____,

_____, _____, _____, _____

STUDENT ASSESSMENT

1 Continue the counting sequence.

   a 25, 30, 35, 40, _____, _____, _____, _____

   b 5, 7, 9, 11, 13, _____, _____, _____, _____

   c 47, 52, 57, 62, _____, _____, _____, _____

   d 13, 15, 17, 19, _____, _____, _____, _____

   e 20, 22, 24, 26, _____, _____, _____, _____

2 Fill in the missing numbers.

   a 3, 6, 9, 12, _____, _____, 21, 24, 27, _____

   b 13, 23, 33, 43, _____, 63, _____, _____, 93, 103

   c 36, 33, _____, 27, _____, 21, _____, 15, 12, _____

   d 1, 4, 7, _____, _____, 16, _____, 22, 25, _____

3 Continue the counting sequence.

   a 89, 79, 69, 59, _____, _____, _____, _____

   b 55, 60, 65, 70, _____, _____, _____, _____

   c 38, 36, 34, 32, _____, _____, _____, _____

   d 24, 34, 44, 54, _____, _____, _____, _____

   e 85, 80, 75, 70, _____, _____, _____, _____

Unit
1
Counting (TRB pp. 22–25)
Whole numbers MA1-4NA applies place value, informally, to count, order, read and represent two- and three-digit numbers

7

How Many Do You Think?

1 a Do you think there are about 5, 25 or 250 faces? _____

b Why do you think that? _____

_____

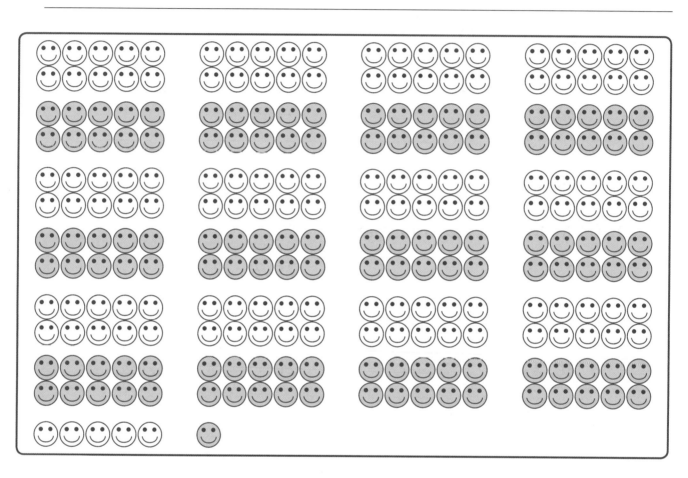

2 How do you think you could count the faces?

_____

3 Count the faces. There are _____ faces.

4 Write a number that is **more than** the number of faces. _____

5 Write a number that is **less than** the number of faces. _____

6 Write a number that is **100 more than** the number of faces. _____

7 Write a number that is **10 less than** the number of faces. _____

Unit 2 Modelling Numbers (TRB pp. 26–29)
Whole numbers MA1-4NA applies place value, informally, to count, order, read and represent two- and three-digit numbers

Modelling with Craft Sticks

You will need: a copy of BLM 4 'Sticks and Bundles', scissors, glue

Look at the 3 numbers.

| 1 | | 4 | | 2 |

1 Make as many 3-digit numbers as you can.

2 Write the **smallest** number in the little box below.

Show this number using the craft sticks from BLM 4.

Paste them in the big box.

3 Write the **largest** number in the little box below.

Show this number using the craft sticks from BLM 4.

Paste them in the big box.

4 Write and show a number that is **more than** your number in Question 2 but **less than** your number in Question 3.

Unit **2** Modelling Numbers (TRB pp. 26–29)
Whole numbers MA1-4NA applies place value, informally, to count, order, read and represent two- and three-digit numbers

9

What Number Could it Be?

You will need: a copy of BLM 6 'Pictures of MAB', scissors, glue

Ali made a number. He used 3 of some blocks, 5 of another kind and 2 of another kind.

Use the pictures from BLM 6 to show 4 different ways that Ali's number could have looked.

a

What number is it? _____

b

What number is it? _____

c

What number is it? _____

d

What number is it? _____

Unit 2 Modelling Numbers (TRB pp. 26–29)
Whole numbers MA1-4NA applies place value, informally, to count, order, read and represent two- and three-digit numbers

DATE:

STUDENT ASSESSMENT

1 a Write the numbers shown.

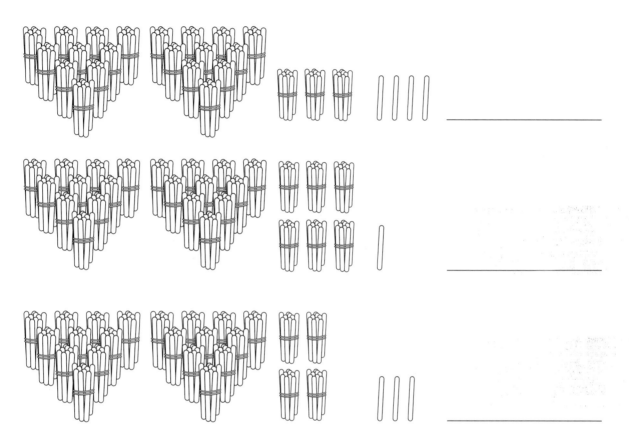

_____

_____

_____

b Draw a circle around the **largest** number.

2 a Write the numbers shown.

_____   _____   _____

b Draw a circle around the **smallest** number.

3 Draw a circle around the **largest** number.

511

Unit
2
Modelling Numbers (TRB pp. 26–29)
Whole numbers MA1-4NA applies place value, informally, to count, order, read and represent two- and three-digit numbers

11

Hands and Feet

You will need: paper, a pencil, scissors, a partner

1 On a sheet of paper, trace around your foot and cut out the footprint shape.

2 Use your hand span and footprint to measure objects in the classroom. Estimate first and then measure with the help of your partner.

	Hand span		Footprint	
	Estimate	Actual measurement	Estimate	Actual measurement
Length of the table				
Height of the chair				
Width of the door				
Your choice				

3 Was it easier to measure with your hand span or your footprint? _____

Why? _____

Unit 3

Length (TRB pp. 30–33)
Length MA1-9MG measures, records, compares and estimates lengths and distances using uniform informal units, metres and centimetres

Measuring in Metres

You will need: a metre ruler or string or tape that is 1 metre long

A metre is about as long as a very long stride.

We can write one metre like this: 1 m.

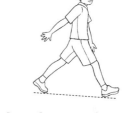

1 Draw something in the classroom that is:

more than a metre	**less than** a metre	**about** a metre

2 How many metres do you estimate it is across

the classroom? ☐

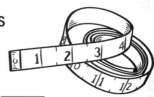

3 How many strides is it across the classroom? ☐

4 Measure how many metres it is across the classroom using your

metre measure. ☐

5 How many of these will fit along your metre measure?

First estimate, then find out.

	Estimate	Actual (measurement)
foot lengths		
paint brushes		
hand spans		

Measuring in Centimetres

You will need: MAB tens

Your finger tip is about one centimetre across.

We can write one centimetre like this: I cm.

1 Draw something in the classroom that is:

more than I cm	**less than** I cm	**about** I cm

A MAB ten is I0 cm long.

10 cm

2 This line is 2 cm long. ▬

How long are these lines? First estimate, then measure using a MAB ten.

Estimate	Actual (measurement)
cm	cm
cm	cm
cm	cm

3 How long are these? First estimate, then measure using a MAB ten.

	Estimate	Actual (measurement)
my pencil		
a pair of scissors		
my foot		
my hand span		

Unit 3

STUDENT ASSESSMENT

You will need: MAB tens

1 Tick ✓ the picture to show what you would use to find the length of the eraser:

My eraser

finger tip widths ☐

foot lengths ☐

hand spans ☐

2 What is the length of the eraser? _____

3 This line is 1 cm long. ▬

How long are these lines? Estimate, then measure using MAB tens.

Estimate	Actual (measurement)	
cm	cm	▬▬▬
cm	cm	▬▬▬▬▬
cm	cm	▬▬▬▬▬▬▬▬▬

4 Tick ✓ the best estimate for the height of a door.

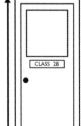

2 centimetres ☐   20 centimetres ☐

2 metres ☐   20 metres ☐

Unit 3 **Length** (TRB pp. 30–33)
Length MA1-9MG measures, records, compares and estimates lengths and distances using uniform informal units, metres and centimetres

15

Arrow Cards

You will need: a set of arrow cards made from a copy of BLM 7 'Arrow Cards 1' and BLM 8 'Arrow Cards 2'

1 Using your arrow cards, make numbers to match the MAB models. Write your numbers beside each model.

a 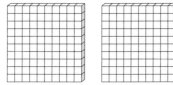 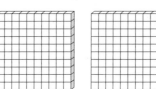 _____

b 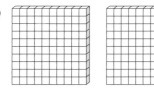 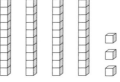 _____

c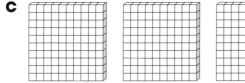

d 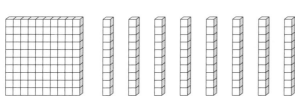 _____

2 Use your remaining cards to make the numbers below. Write your number in the box.

a A number that **is more than 700** but has **no tens**

b A number that is **less than 700** but has **no ones**

c A number that is **between 700** and **900**

Spike Abacus

1 Write the number that is shown on each abacus.

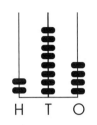
H T O
_____

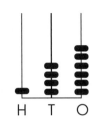
H T O
_____

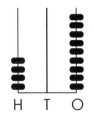
H T O
_____

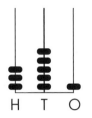
H T O
_____

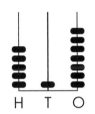
H T O
_____

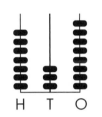
H T O
_____

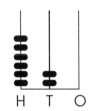
H T O
_____

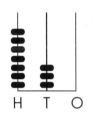
H T O
_____

2 Show the following numbers on the abacus.

H T O

562

H T O

905

H T O

227

H T O

69

H T O

380

H T O

631

H T O

713

H T O

454

3 Show the following numbers on the abacus.

a 10 less than 256

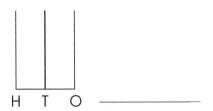
H T O _____

b 100 more than 752

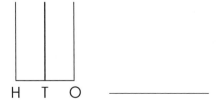
H T O _____

Unit 4 **Numbers up to 1000** (TRB pp. 34–37)
Whole numbers MA1-4NA applies place value, informally, to count, order, read and represent two- and three-digit numbers

17

Climbing the Ladder

You will need: a partner, 3 dice

1 Take turns with your partner to roll the dice and make a 3-digit number. Write your number on a rung of the first ladder. The numbers need to get **bigger** as you go up the ladder.

2 Once you have written a number, it must stay on that rung.

3 Keep rolling the dice and placing your numbers until each player has filled their ladder.

4 You score 1 point if your numbers are in order from **smallest** to **largest**.

5 The person with the most points after 8 rounds wins.

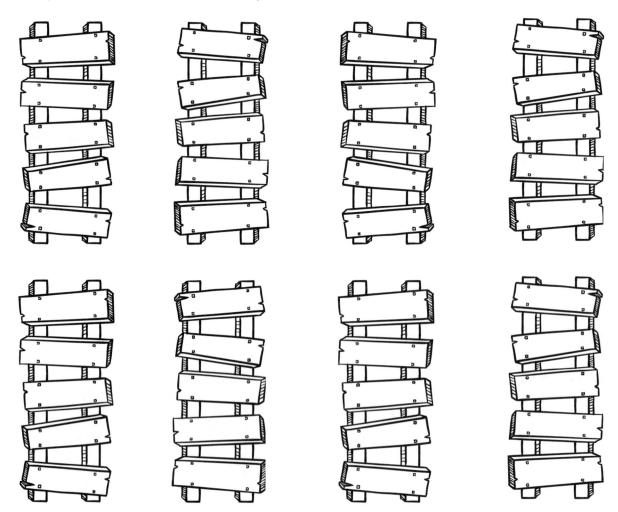

4 STUDENT ASSESSMENT

You will need: a set of arrow cards made from a copy of BLM 7 'Arrow Cards 1' and BLM 8 'Arrow Cards 2', a dice

1 Use the arrow cards to make numbers to match the MAB below. Write your numbers below each model.

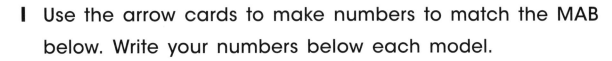

_____   _____

2 Write the numbers that are shown on the abacus.

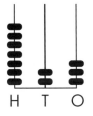   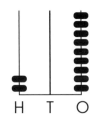   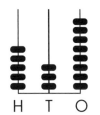   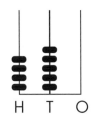

_____   _____   _____   _____

3 Show the following numbers on the abacus.

562   147   302   415

4 a Roll a dice 3 times. Write each number in a box.

b Use the numbers to write as many 3-digit numbers as you can.

_____

c Order your numbers from **smallest** to **largest**.

_____

Unit
4
Numbers up to 1000 (TRB pp. 34–37)
Whole numbers MA1-4NA applies place value, informally, to count, order, read and represent two- and three-digit numbers

19

What's the Total?

You will need: 2 dice

1 Roll the dice. Write each number in a box. Circle the **larger** number. Now count on from the larger number to find the total.

◻ and ◻ is _____

2 Try again and again!

◻ and ◻ is _____   ◻ and ◻ is _____

◻ and ◻ is _____   ◻ and ◻ is _____

◻ and ◻ is _____   ◻ and ◻ is _____

◻ and ◻ is _____   ◻ and ◻ is _____

3 a What was the **largest** total you made? _____

  b Could you have made a larger total? _____ Why? Why not?

_____

4 a What was the **smallest** total you made? _____

  b Could you have made a smaller total? _____ Why? Why not?

_____

Double Everything

1 Imagine that you have a pot like the Haktaks (in *Two of Everything*), which doubles everything that is dropped into it. Draw what would happen **after** you dropped in:

double 1 is _____

double 5 is _____

double 3 is _____

double 6 is _____

double 4 is _____

double 2 is _____

2 Circle the doubles problems, and then work them out.

4 + 3 =	4 + 4 =	6 + 3 =	2 + 3 =
2 + 2 =	1 + 6 =	3 + 2 =	1 + 1 =
6 + 6 =	5 + 4 =	3 + 3 =	4 + 5 =

Unit 5 **Strategies for Addition** (TRB pp. 38–41)
Addition and subtraction MA1-5NA uses a range of strategies and informal recording methods for addition and subtraction involving one- and two-digit numbers

21

Facts for Ten

You will need: coloured pencils

1 Colour the first square yellow and the remaining squares blue.

1 and _____ make 10   **and**   _____ and 1 make 10

2 Colour the first 2 squares yellow and the remaining squares blue.

2 and _____ make 10   **and**   _____ and 2 make 10

3 Colour the first 3 squares yellow and the remaining squares blue.

3 and _____ make 10   **and**   _____ and 3 make 10

4 Colour the first 4 squares yellow and the remaining squares blue.

4 and _____ make 10   **and**   _____ and 4 make 10

5 Colour the first 5 squares yellow and the remaining squares blue.

5 and _____ make 10

6 Circle all the problems that can be solved with **facts for ten**.
Solve the remaining problems using counting on.

6 + 3 =	5 + 5 =	9 + 1 =	8 + 2 =
2 + 8 =	2 + 3 =	10 + 1 =	3 + 8 =
4 + 3 =	3 + 7 =	7 + 3 =	2 + 4 =
1 + 9 =	2 + 5 =	5 + 3 =	3 + 4 =

Unit **5** **Strategies for Addition** (TRB pp. 38–41)
Addition and subtraction MA1-5NA uses a range of strategies and informal recording methods for addition and subtraction involving one- and two-digit numbers

STUDENT ASSESSMENT

1 Complete the following.

3 and 3 is _____   6 and 6 is _____

2 and 2 is _____   5 and 5 is _____

1 and 1 is _____   4 and 4 is _____

2 Complete the dominoes so that their sum (total) is 10.

  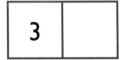 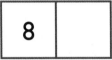

3 a Solve the following problems.

7 + 4 =   2 + 5 =

6 + 1 =   3 + 4 =

2 + 9 =   8 + 1 =

3 + 5 =   7 + 2 =

6 + 3 =   2 + 4 =

5 + 4 =   8 + 3 =

b Explain how you solved the problems.

_____

_____

Unit
5
Strategies for Addition (TRB pp. 38–41)
Addition and subtraction MA1-5NA uses a range of strategies and informal recording methods for addition and subtraction
involving one- and two-digit numbers

23

How Could They Be Partitioned?

You will need: coloured pencils

1 Jack has 6 cubes; some are yellow and some are green.
Draw 2 ways the cubes could look.

6 can be partitioned into _____ and _____ **or** _____ and _____

2 Jess has 8 cubes; some are blue and some are yellow.
Draw 2 ways the cubes could look.

8 can be partitioned into _____ and _____ **or** _____ and _____

3 Ji has 5 cubes; some are red and some are purple.
Draw 2 ways the cubes could look.

5 can be partitioned into _____ and _____ **or** _____ and _____

4 Jeb has 7 cubes; some are brown and some are green.
Draw 2 ways the cubes could look.

7 can be partitioned into _____ and _____ **or** _____ and _____

5 Jan has 9 cubes; some are blue and some are purple.
Draw 2 ways the cubes could look.

9 can be partitioned into _____ and _____ **or** _____ and _____ .

More Strategies for Addition (TRB pp. 42–45)
Addition and subtraction MA1-5NA uses a range of strategies and informal recording methods for addition and subtraction involving one- and two-digit numbers

Building to 10 Strategy

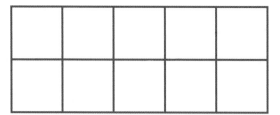

You will need: 10 counters

1 Place 7 counters on the ten frame.

How many more do you need to make 10? _____

Use what you have learned to solve the problems.

One has been done for you.

7 + 6 =   7 + 9 =   7 + 4 =

7 + 3 + 3 =

10 + 3 = 13

2 Place 9 counters on the ten frame.

How many more do you need to make 10? _____

Use what you have learned to solve the problems.

9 + 8 =   9 + 3 =   9 + 5 =

3 Place 5 counters on the ten frame.

How many more do you need to make 10? _____

Use what you have learned to solve the problems.

5 + 8 =   5 + 9 =   5 + 6 =

Unit 6

More Strategies for Addition (TRB pp. 42–45)
Addition and subtraction MA1-5NA uses a range of strategies and informal recording methods for addition and subtraction involving one- and two-digit numbers

25

Ten More

You will need: coloured pencils

1	2	3	4	5	6	7	8	9	10
11	12	13	14	15	16	17	18	19	20
21	22	23	24	25	26	27	28	29	30
31	32	33	34	35	36	37	38	39	40
41	42	43	44	45	46	47	48	49	50
51	52	53	54	55	56	57	58	59	60
61	62	63	64	65	66	67	68	69	70
71	72	73	74	75	76	77	78	79	80
81	82	83	84	85	86	87	88	89	90
91	92	93	94	95	96	97	98	99	100

1 Find 37 on the 100 chart and colour the number.

To answer 37 + 10 = _____, count on 10 and colour the number.

2 Find 52 on the 100 chart and colour the number.

To answer 52 + 10 = _____, count on 10 and colour the number.

3 What do you notice about the second numbers that you coloured in Questions 1 and 2?

_____

_____

4 **a** Solve the following problems.

84 + 10 =   24 + 10 =   79 + 10 =   46 + 10 =

58 + 10 =   67 + 10 =   33 + 10 =   15 + 10 =

8 + 10 =   43 + 10 =   26 + 10 =   74 + 10 =

31 + 10 =   86 + 10 =   54 + 10 =   63 + 10 =

b Use the 100 chart to check your answers.
Colour the boxes that have the correct answers.

STUDENT ASSESSMENT

DATE:

1 Write some number sentences to show how:

7 can be partitioned

9 can be partitioned

2 Use the **building to 10 strategy** to solve the problems.

7 + 5 =   8 + 7 =   6 + 7 =

3 Use the **adding 10 strategy** to solve the problems.

34 + 10 =   18 + 10 =   25 + 10 =

4 What is the total shown on the dice? _____

Explain the strategy that you used.

_____

Unit 6 **More Strategies for Addition** (TRB pp. 42–45)
Addition and subtraction MA1-5NA uses a range of strategies and informal recording methods for addition and subtraction involving one- and two-digit numbers

27

Paper Folding

You will need: small paper squares (kinder squares)

1 Use paper squares to find different ways to fold them in **half**.

Draw some different ways below.

2 Now try to fold the paper into 4 equal parts to make **quarters**.

Draw 2 different ways below.

3 Can you fold the paper into 8 equal parts to make **eighths**?

Draw 2 different ways to fold the paper into eighths.

4 If your paper square was a slice of bread, which fraction would give you the **biggest** piece of bread? _____

Dividing Shapes

1 Has the shape been divided
   into **quarters**? _____

   How do you know?

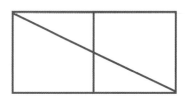

2 Has the shape been divided
   into **quarters**? _____

   How do you know?

3 Has the shape been divided
   into **quarters**? _____

   How do you know?

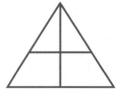

4 Has the shape been divided
   into **quarters**? _____

   How do you know?

5 Divide each shape into **quarters**. Colour **one quarter** of each shape.

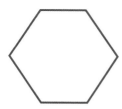 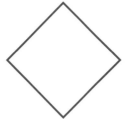 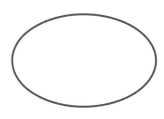

6 Divide each shape into **eighths**. Colour **one eighth** of each shape.

 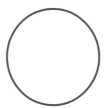

Unit **7** **Fractions** (TRB pp. 46–49)
Fractions and decimals MA1-7NA represents and models halves, quarters and eighths

29

Picture, Word, Symbol

1 Complete the table. One row has been done for you.

	Write the fraction word	Write the fraction symbol
Cut the pizza into 2 equal parts and colour 1 part.	half	$\dfrac{1}{2}$
Cut the pizza into 4 equal parts and colour 1 part.		
Cut the pizza into 8 equal parts and colour 1 part.		

2 If Ray ate this much of a pizza:

   a What fraction did he eat? _____

   b Is there another way of writing or showing this amount? _____

3 If Rita ate this much of a pizza:

   a What fraction did she eat? _____

   b Is there another way of writing or showing this amount? _____

4 Ling ate **one half** of a pizza and Bella ate **two quarters**.

   Who ate the most pizza? Explain your answer.

DATE:

STUDENT ASSESSMENT

I Ted has tried to cut the shapes in half, by cutting along the lines that are shown.

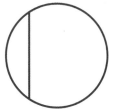 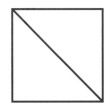 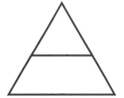 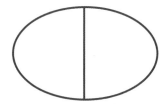

What advice would you give him?

_____

_____

2 Divide each shape into quarters.

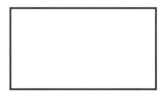 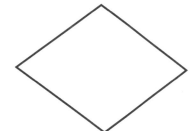 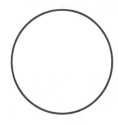

3 Kai wants to eat $\frac{1}{8}$ of the pizza. Colour how much Kai wants to eat.

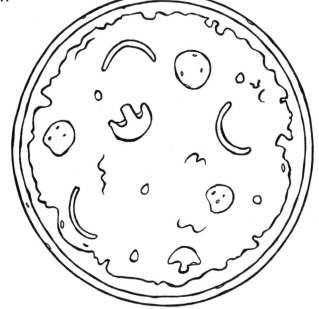

Unit 7   **Fractions** (TRB pp. 46–49)
Fractions and decimals MA1-7NA represents and models halves, quarters and eighths

31

Shapes

You will need: a copy of BLM 17 'Shapes', scissors, glue, a partner

1 Cut out the shapes from BLM 17. Paste each shape in the correct place in the first column. Fill in the remaining columns.

Shape	How many edges?	How many corners?
square		
triangle		
rectangle		
rhombus		
kite		

2 Draw a mystery shape and describe it to your partner.

3 Draw your partner's mystery shape.

Unit 8 **Transformation of 2D Shapes** (TRB pp. 50–53)
Two-dimensional space MA1-15MG manipulates, sorts, represents, describes and explores two-dimensional shapes, including quadrilaterals, pentagons, hexagons and octagons

Flips and Slides

1 Describe the slides up, down and across to move the shape in each grid from ⬛ to ⬜ .

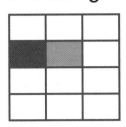 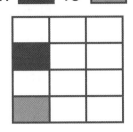 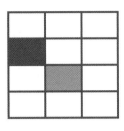 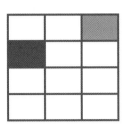

_____ _____ _____ _____

_____ _____ _____ _____

2 Draw what you think each shape would look like if it were **flipped horizontally**. Look at the picture of the triangle flip to help you.

   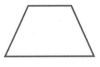

3 Draw what you think each shape would look like if it were **flipped vertically**. Look at the picture of the triangle flip to help you.

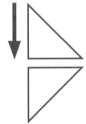  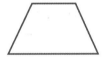

4 Describe how you would slide or flip each shape to fit it into the pegboard.

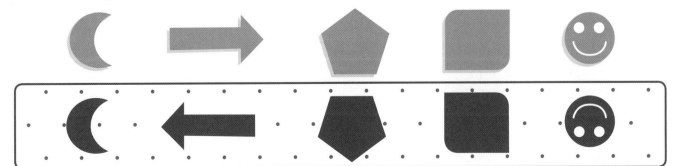

Unit 8
Transformation of 2D Shapes (TRB pp. 50–53)
Two-dimensional space MA1-15MG manipulates, sorts, represents, describes and explores two-dimensional shapes, including quadrilaterals, pentagons, hexagons and octagons

33

Quarter and Half Turns

You will need: a copy of BLM 17 'Shapes', scissors, glue

1 a Cut out the shapes from BLM 17. Paste each shape in the correct place in the first column of the table.

b Draw what the shape would look like after **a quarter turn**.

c Draw what the shape would look like after **a half turn**.

Shape	Quarter turn	Half turn
rectangle		
triangle		
kite		
rhombus		

2 Finish the pattern.

Unit 8

STUDENT ASSESSMENT

1 Colour the shape that is a rectangle.

How do you know that it is a rectangle?

_____

2 Draw a line to match each shape to its name.

   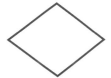

rhombus   kite   triangle   rectangle   circle

3 Draw what the shape will look like when you:

a slide it up   **b** slide it across   **c** slide it down

4 Draw what the shape will look like if it is **flipped horizontally**.

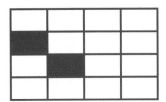

5 Write how you would turn the shape to make it fit into the jigsaw.

 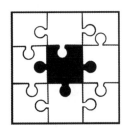

_____

6 To finish the pattern, write how you would turn the kite.

_____

Unit 8

Transformation of 2D Shapes (TRB pp. 50–53)
Two-dimensional space MA1-15MG manipulates, sorts, represents, describes and explores two-dimensional shapes, including quadrilaterals, pentagons, hexagons and octagons

35

Adding on a Ten Frame

You will need: counters

1 Use counters and the ten frames to help you solve the problems.

17	19	14
+ 8	+ 2	+ 11

<table>
<tr><td colspan="5"></td></tr>
<tr><td></td><td></td><td></td><td></td><td></td></tr>
<tr><td></td><td></td><td></td><td></td><td></td></tr>
</table>

13	12	12
+ 5	+ 9	+ 16

11	3	15
+ 6	+ 16	+ 12

4	18	19
+ 15	+ 5	+ 6

2 Jeb rolled 6, 3 and 8 on three 10-sided dice. Kai rolled 5, 9 and 2. When the numbers were added together, who had the **highest** score? Show your working out.

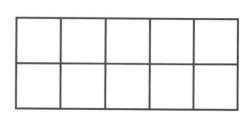

3 Frankie rolled 4, 3 and 8 on three 10-sided dice. Then Ali rolled the dice. When Ali's numbers were added together, he scored the same as Frankie. What might Ali have rolled? Show your working out.

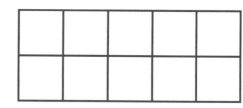

Solving Addition Problems (TRB pp. 54–57)
Addition and subtraction MA1-5NA uses a range of strategies and informal recording methods for addition and subtraction involving one- and two-digit numbers

Empty Number Lines

Use the empty number lines to solve the following problems.
One has been started for you.

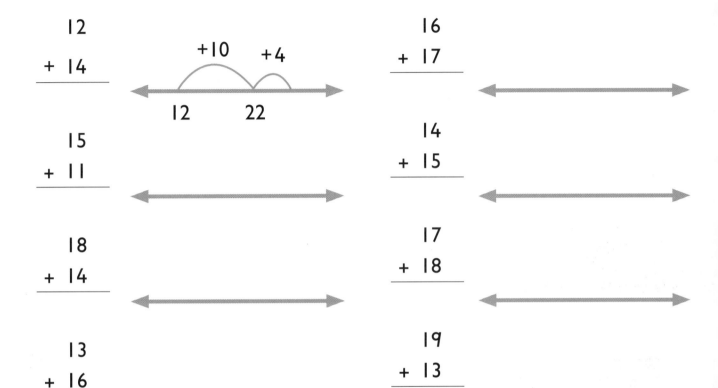

12
+ 14
_____

16
+ 17
_____

15
+ 11
_____

14
+ 15
_____

18
+ 14
_____

17
+ 18
_____

13
+ 16
_____

19
+ 13
_____

11
+ 12
_____

15
+ 16
_____

Challenge!

Use the empty number lines to solve the following problems.
One has been started for you.

14
+
_____
27

18
+
_____
35

Unit 9

Solving Addition Problems (TRB pp. 54–57)
Addition and subtraction MA1-5NA uses a range of strategies and informal recording methods for addition and subtraction involving one- and two-digit numbers

37

Is it Balanced?

You will need: a partner, a number balance

1 Work with a partner and use a number balance to solve the problems. Write an equivalent statement. One has been done for you.

$9 + 4 = 8 + 5$

$6 + 7 = \boxed{} + \boxed{}$

$4 + 8 = \boxed{} + \boxed{}$

$9 + 9 = \boxed{} + \boxed{}$

$6 + 9 = \boxed{} + \boxed{}$

$3 + 7 = \boxed{} + \boxed{}$

$8 + 5 = \boxed{} + \boxed{}$

$7 + 10 = \boxed{} + \boxed{}$

$2 + 9 = \boxed{} + \boxed{}$

$7 + 6 = \boxed{} + \boxed{}$

2 If each side of the balance was equal to 16, draw what you think the balance could look like.

Solving Addition Problems (TRB pp. 54–57)
Addition and subtraction MA1-5NA uses a range of strategies and informal recording methods for addition and subtraction involving one- and two-digit numbers

STUDENT ASSESSMENT

DATE:

1 Imagine that Don has 12 counters on 2 ten frames. He then adds 7 more counters. How many does he have altogether? _____

2 Use the empty number lines to work out how many students are in each class.

Class	Boys	Girls	Total
Year 1	16	13	
Year 2	15	11	
Year 3	12	13	
Year 4	14	14	
Year 5	17	12	
Year 6	11	15	

3 Mark the numbers on the number balance and complete the equivalence statements.

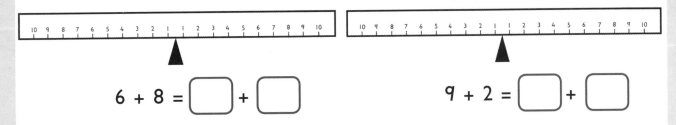

6 + 8 = ☐ + ☐

9 + 2 = ☐ + ☐

Unit
9
Solving Addition Problems (TRB pp. 54–57)
Addition and subtraction MA1-5NA uses a range of strategies and informal recording methods for addition and subtraction involving one- and two-digit numbers

39

Pattern Block Pictures

You will need: a copy of BLM 21 'Pattern Blocks', scissors, glue, a partner

1 Cut out the pattern blocks from BLM 21. Make a picture with the blocks and paste your picture below. Make sure your partner cannot see your picture.

2 Have your partner get another sheet of paper. Tell them where to draw the shapes so that they make a picture that is **exactly** the same as yours.

3 Now listen to your partner and draw what they tell you on another sheet of paper.

4 Is your partner's picture the same as yours? _____

How is it different?

_____

Ebony's Run

One day, Ebony got up early to go for a run. She walked out of her **big house** and ran down to the **end of her street** where there was a **big forest**. There were **2 paths** into the forest, and she took the path on the right. She ran until she was in the middle of the forest, where she found a **small cave**. She heard a growl coming from the cave and quickly ran past the cave and further into the forest. Ebony thought she was lost until she came to a **road**. She turned left into the road and followed it all the way to **her street**. She ran along her street until she got **home**.

Draw a map of where Ebony ran. Remember to include all of the places mentioned in the story.

Unit 10 **Position** (TRB pp. 58–61)
Position MA1-16MG represents and describes the positions of objects in everyday situations and on maps

41

Our School

1 List the things that are part of your school buildings.

_____

2 List the things that are outside the buildings but part of the school.

_____

3 Draw a map of your school. Make sure you include all the things on your lists. Remember that a map is a bird's-eye view.

4 Is there anything else you can add to the map of your school?

_____

STUDENT ASSESSMENT

1 Write how to get from your classroom to the office.

_____

_____

2 What could the following be? You are looking at them from a bird's-eye view.

[rectangle]   [circle]

_____   _____

_____   _____

3 Draw a map of your classroom.

Unit
10
Position (TRB pp. 58–61)
Position MA1-16MG represents and describes the positions of objects in everyday situations and on maps

43

Solve by Counting Back

You will need: a six-sided dice, a sheet of paper

0	1	2	3	4	5	6	7	8	9	10	11	12	13	14	15	16	17	18	19	20

1 Roll the dice and write the number in a blank space. Use the number track above to help you **count back** to solve the problems. Write your answers in the boxes.

$8 -$ _____ $=$ ☐ $12 -$ _____ $=$ ☐

$9 -$ _____ $=$ ☐ $11 -$ _____ $=$ ☐

$14 -$ _____ $=$ ☐ $10 -$ _____ $=$ ☐

$7 -$ _____ $=$ ☐ $15 -$ _____ $=$ ☐

$9 -$ _____ $=$ ☐ $7 -$ _____ $=$ ☐

$13 -$ _____ $=$ ☐ $10 -$ _____ $=$ ☐

$13 -$ _____ $=$ ☐ $11 -$ _____ $=$ ☐

$8 -$ _____ $=$ ☐ $14 -$ _____ $=$ ☐

2 Cover the number track above. Try to work out the problems by counting back in your head.

$8 - 3 =$ $11 - 2 =$ $4 - 2 =$

$12 - 3 =$ $10 - 4 =$ $7 - 3 =$

$9 - 4 =$ $6 - 3 =$ $11 - 3 =$

$9 - 3 =$ $5 - 2 =$ $10 - 2 =$

Solve by Counting On

| 0 | 1 | 2 | 3 | 4 | 5 | 6 | 7 | 8 | 9 | 10 | 11 | 12 | 13 | 14 | 15 | 16 | 17 | 18 | 19 | 20 |

1 Use the number track above to help you **count on** to solve the problems.

$19 - 17 =$ $18 - 15 =$ $17 - 15 =$ $18 - 14 =$

$15 - 12 =$ $16 - 11 =$ $16 - 13 =$ $18 - 16 =$

$18 - 13 =$ $19 - 15 =$ $17 - 14 =$ $16 - 12 =$

$17 - 13 =$ $19 - 16 =$ $15 - 9 =$ $19 - 14 =$

$15 - 11 =$ $15 - 13 =$ $16 - 14 =$ $17 - 12 =$

2 It is best to use the **counting on strategy** when the numbers in the problem are close to one another. Circle the problems below that you would solve by using the counting on strategy.

$18 - 14 =$ $20 - 16 =$ $17 - 14 =$ $17 - 13 =$

$16 - 5 =$ $17 - 3 =$ $16 - 15 =$ $17 - 4 =$

$19 - 17 =$ $18 - 15 =$ $20 - 18 =$ $18 - 4 =$

$20 - 8 =$ $18 - 5 =$ $19 - 7 =$ $20 - 6 =$

$14 - 2 =$ $19 - 6 =$ $14 - 12 =$ $19 - 16 =$

3 Solve the problems that you circled in Question 2.

Unit **11** **Strategies for Subtraction** (TRB pp. 62–65)
Addition and subtraction MA1-5NA uses a range of strategies and informal recording methods for addition and subtraction involving one- and two-digit numbers

45

Lowest Number Scores

You will need: a 10-sided dice, a partner

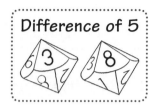

Difference of 5

1 Fill in your name and your partner's name in the table.

2 Roll the dice twice. Record the numbers in the table. Now find and record the **difference** between the 2 numbers.

3 Have your partner roll the dice. Then record their numbers and the difference between them.

4 In each row, circle the difference that is the **lowest**.

5 Continue until all the rows are filled in.

6 The person with the most circles wins.

Your name:			Your partner's name:		
1st number	2nd number	Difference	1st number	2nd number	Difference

Unit 11

STUDENT ASSESSMENT

1 Solve the following problems by using the **counting back strategy**.

16 – 4 =   19 – 7 =   15 – 6 =

12 – 3 =   17 – 4 =   13 – 5 =

2 Solve the following problems by using the **counting on strategy**.

18 – 15 =   16 – 14 =   20 – 17 =

15 – 12 =   19 – 17 =   17 – 14 =

3 Find the difference between the pairs of numbers.

8 and 14   16 and 3   7 and 9

13 and 7   5 and 12   15 and 13

4 Solve the following problem.

16 – 11 =

Explain how you solved it.

_____

Unit 11 **Strategies for Subtraction** (TRB pp. 62–65)
Addition and subtraction MA1-5NA uses a range of strategies and informal recording methods for addition and subtraction involving one- and two-digit numbers

47

Subtracting on a Ten Frame

DATE:

You will need: counters

1 Use counters and the ten frames to help you solve the problems.

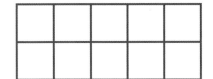 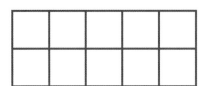 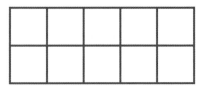

29 – 8 =	30 – 17 =	18 – 5 =	19 – 8 =
16 – 11 =	26 – 9 =	22 – 7 =	17 – 13 =
20 – 12 =	15 – 4 =	25 – 12 =	21 – 5 =
24 – 9 =	21 – 8 =	26 – 8 =	15 – 7 =
19 – 12 =	28 – 13 =	23 – 6 =	27 – 12 =

2 Work out how many students are present in each class.

Class	Total number of students in class	Number of students absent	Number of students present
1A	21	4	
1B	23	3	
1C	22	6	
2A	25	1	
2B	24	5	
2C	26	2	

Unbundle or Not?

You will need: craft sticks in singles and bundles of 10, or a copy of BLM 4 'Sticks and Bundles'

1 Use craft sticks to work out the answers to the problems.

$$24 - 17$$

$$25 - 13$$

2 a Circle the problem in Question 1 where you needed to **unbundle** a bundle of 10 craft sticks.

b Why did you need to unbundle the craft sticks for that problem? _____

_____

3 Circle the problems where you will need to **unbundle** a bundle of 10 craft sticks. Solve **only** those problems.

$$27 - 14$$ $$23 - 12$$ $$22 - 11$$ $$29 - 8$$

$$22 - 15$$ $$25 - 7$$ $$27 - 18$$ $$21 - 8$$

$$24 - 8$$ $$29 - 16$$ $$28 - 15$$ $$26 - 17$$

$$21 - 11$$ $$26 - 12$$ $$24 - 12$$ $$23 - 14$$

Unit 12 **Subtraction** (TRB pp. 66–69)
Addition and subtraction MA1-5NA uses a range of strategies and informal recording methods for addition and subtraction involving one- and two-digit numbers

49

Using MAB for Subtraction

You will need: MAB or a copy of BLM 6 'Pictures of MAB'

1 Solve the problems using MAB. Circle the problems where you need to **partition a ten into 10 ones**, to be able to take away the ones.

27 − 15	33 − 22	42 − 16	39 − 25
48 − 33	25 − 18	46 − 36	24 − 9
37 − 17	43 − 20	27 − 13	33 − 16
45 − 13	26 − 14	41 − 19	30 − 18

2 Amina had $36. She bought a puzzle for $14.
Write the number sentence below. Use MAB to help you find out how much money Amina has left.

3 Asad saved up $18. He wants to buy a game that costs $43.
How much more does he need to save?
Write the number sentence below. Use MAB to help you solve the problem.

Subtraction (TRB pp. 66–69)
Addition and subtraction MA1-5NA uses a range of strategies and informal recording methods for addition and subtraction involving one- and two-digit numbers

Unit 12

STUDENT ASSESSMENT

You will need: counters, single craft sticks and bundles of 10

1 Use counters and the ten frames to solve this problem.

Jo put 25 counters on her ten frames. Vin knocked off 17.
How many did Jo have left? _____

2 Use single craft sticks and bundles of 10 to solve the
problems.

29	23	26
− 16	− 9	− 18

3 Imagine you are using MAB to solve the problems below.
Circle the problems where you would need to **partition
a ten into 10 ones** to get your answer.

24	30	27	26
− 13	− 13	− 19	− 15

21	25	23	22
− 14	− 12	− 16	− 11

4 Solve **only** the problems that you circled in Question 3.

Unit 12 **Subtraction** (TRB pp. 66–69)
Addition and subtraction MA1-5NA uses a range of strategies and informal recording methods for addition and
subtraction involving one- and two-digit numbers

51

Heavy or Light?

You will need: 5 classroom objects, balance scales

1 Place your objects on the balance scales. Draw the objects on the scales below and complete the statements.

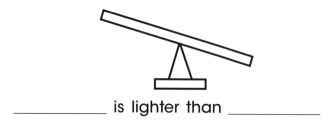

_____ is lighter than _____

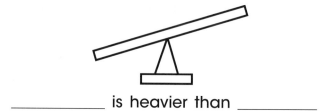

_____ is heavier than _____

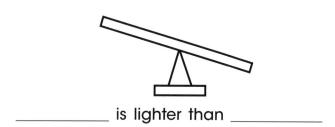

_____ is lighter than _____

_____ is heavier than _____

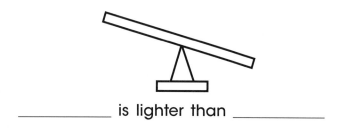

_____ is lighter than _____

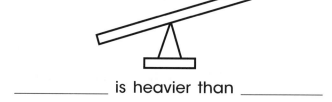

_____ is heavier than _____

2 Put objects on each side of the balance scales to make it level. You might need to use a few things on one side. Draw the objects and complete the statement below.

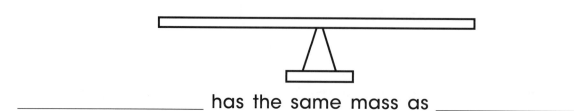

_____ has the same mass as _____

Mass (TRB pp. 70–73)
Mass MA1-12MG measures, records, compares and estimates the masses of objects using uniform informal units

Ordering Objects

You will need: a partner, balance scales, a pencil case, a student book, a drink bottle, a library book, a shoe (or any 5 classroom objects)

1 Estimate the weight of each object from the **lightest** (I) through to the **heaviest** (5). Draw the objects in order below.

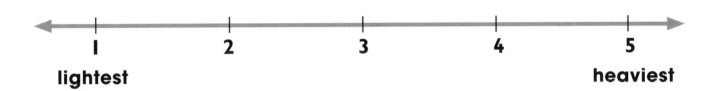

lightest   **heaviest**

2 Work with a partner and use the balance scales to find the weight of the objects. Draw your findings below.

lightest   **heaviest**

3 Explain how you decided on the order of your objects in Question I. _____

_____

_____

4 What else did you discover?

_____

_____

Unit **13** **Mass** (TRB pp. 70–73)
Mass MA1-12MG measures, records, compares and estimates the masses of objects using uniform informal units

53

How Many?

You will need: blocks (or other uniform units such marbles), balance scales, a pencil case, a student book, a glue stick, a drink bottle

1 Estimate how many blocks you will need to balance each of the objects. Write your estimate in the table.

	Estimate	Number of blocks needed to balance
pencil case		
student book		
glue stick		
drink bottle		
Draw something you would like to measure.		

2 Using the blocks and balance scales, weigh each object. Write your results in the table.

3 What strategies did you use to estimate how many blocks it would take to balance each object?

_____

_____

STUDENT ASSESSMENT

Unit 13

You will need: 3 classroom objects, balance scales, blocks

1 Draw how a 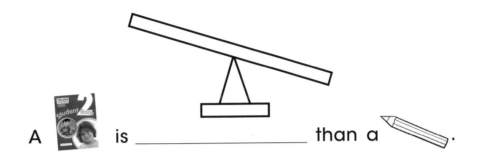 and a ✏️ would look on the balance scales. Complete the statement below.

A 📖 is _____ than a ✏️.

2 Draw how a 🩹glue stick and a 🍶 would look on the balance scales. Complete the statement below.

A 🩹glue stick is _____ than a 🍶.

3 Use balance scales to order your 3 classroom objects from **heaviest** to **lightest**. Draw them in the boxes.

heaviest   **lightest**

4 Use balance scales to find how many blocks are needed to balance your student book. _____

Unit 13 **Mass** (TRB pp. 70–73)
Mass MA1-12MG measures, records, compares and estimates the masses of objects using uniform informal units

55

Clocks

1 Draw the hands on each clock to show the time.

half-past 12

11:30

8 o'clock

12:00

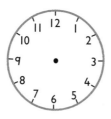

half-past 7

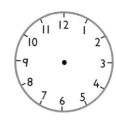

5:30

2 Write, in words, the time shown on each clock.

_____

_____

_____

_____

_____

_____

_____

_____

3 a What is your favourite time of the day? Show this time on both clocks.

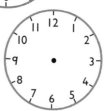

b Why is this your favourite time?

_____

4 Draw something you might do at 7:30 in the morning.

Telling the Time (TRB pp. 74–77)
Time MA1-13MG describes, compares and orders durations of events, and reads half- and quarter-hour time

Show the Time

1 Complete the table. The first row has been done for you.

Written time	Analogue clock	Digital clock
half-past 7		0 7 : 3 0
		:
		0 1 : 3 0
		:
		1 2 : 0 0
quarter-past 8		:
		:
		0 4 : 1 5

2 What time do you go to bed?

Show this time on both clocks.

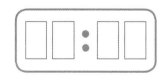

Unit **14** **Telling the Time** (TRB pp. 74–77)
Time MA1-13MG describes, compares and orders durations of events, and reads half- and quarter-hour time

57

Your Day

1 What time do you get up in the morning?
Draw hands on the clock to show the time.

2 What do you do at this time today?

_____

_____

3 What do you do at this time today?

_____

_____

4 When do you eat dinner?
Show the time on the clock.

5 What is your favourite time **today**?
Draw hands on the clock to show the time.

6 a Write 4 times on the clocks below.

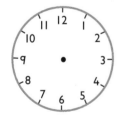   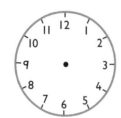   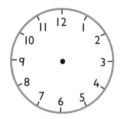   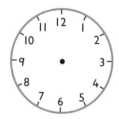

b Circle the clock that shows your **favourite** time.

c Explain why it is your favourite time.

_____

_____

Telling the Time (TRB pp. 74–77)
Time MA1-13MG describes, compares and orders durations of events, and reads half- and quarter-hour time

DATE:

STUDENT ASSESSMENT

1 Write the digital time for each clock.

2 Show the digital times on the clocks.

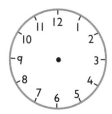 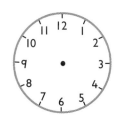  

| 0 2 : 1 5 | 0 4 : 0 0 | 1 1 : 4 5 | 0 8 : 3 0 |

3 Show the times on the clocks in order from **earliest** to **latest**.

| 9 o'clock | quarter-to 8 | half-past 7 | quarter-past 12 |

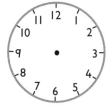 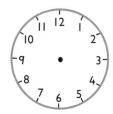  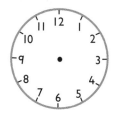

Unit
14
Telling the Time (TRB pp. 74–77)
Time MA1-13MG describes, compares and orders durations of events, and reads half- and quarter-hour time

59

Subtracting 10

You will need: MAB or a copy of BLM 6 'Pictures of MAB'

1 a Use MAB to help you work out the problems.

56	24	72	39
− 10	− 10	− 10	− 10

83	17	41	67
− 10	− 10	− 10	− 10

b What do you notice about the answers?

_____

2 Try to work out the following problems **without** using MAB.

94	66	35	80
− 10	− 10	− 10	− 10

3 Complete the problems.

76	43	92	54
− 20	− 30	− 20	− 30

31	61	86	29
− 20	− 10	− 30	− 20

4 At a huge sale, every item costs $10 less than the price on the tag. Work out the new sale price for each item.

_____   _____   _____   _____

Difference of 9

1 Complete the first problem in each pair. Then use the answer to help you solve the second problem in the pair.

34	34	85	85
− 10	− 9	− 10	− 9

27	27	61	61
− 10	− 9	− 10	− 9

46	46	93	93
− 10	− 9	− 10	− 9

56	56	77	77
− 10	− 9	− 10	− 9

75	75	21	21
− 10	− 9	− 10	− 9

63	63	84	84
− 10	− 9	− 10	− 9

2 Circle the pairs of numbers next to each other that have a difference of 9. Use the answers from the problems in Question 1 to help you. Pairs can be found horizontally, vertically and diagonally. One has been done for you.

21	34	11	63	12	13	76	14
12	15	25	16	54	17	85	56
18	77	68	19	84	20	21	47
22	23	24	93	25	26	27	28
46	29	30	75	66	61	52	31
37	32	27	18	33	34	35	84
36	38	38	40	41	42	52	75

Unit 15 **More About Subtraction** (TRB pp. 78–81)
Addition and subtraction MA1-5NA uses a range of strategies and informal recording methods for addition and subtraction involving one- and two-digit numbers

61

Number Lines Can Help

I Use the number lines to help you solve the problems.

27
− 14

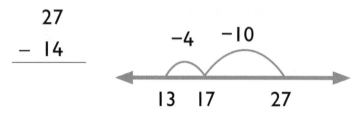

37
− 15

56
− 19

49
− 16

28
− 11

76
− 13

42
− 14

Challenge!

2 a Use the number line to help you solve the following problem.

38
−
23

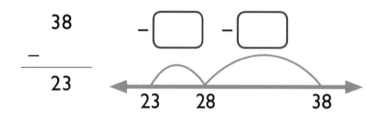

b Try to solve the following problems in the same way.

46
−
35

29
−
14

Unit 15 STUDENT ASSESSMENT

1 Take away 10 from the following numbers.

74 _____   56 _____   92 _____

37 _____   19 _____   41 _____

28 _____   63 _____   85 _____

How did you take away 10?

_____

2 Take away 9 from the following numbers.

51 _____   37 _____   14 _____

85 _____   43 _____   68 _____

23 _____   96 _____   72 _____

What strategy did you use to take away 9?

_____

3 Use the empty number line to solve the following problems.

```
  35
– 14
_____        ←————————→
```

```
  78
– 17
_____        ←————————→
```

```
  56
– 12
_____        ←————————→
```

```
  29
– 13
_____        ←————————→
```

Unit 15 **More About Subtraction** (TRB pp. 78–81)
Addition and subtraction MA1-5NA uses a range of strategies and informal recording methods for addition and subtraction involving one- and two-digit numbers

63

Ten Frames Can Help

You will need: 20 counters in two different colours

1 Place 6 counters of one colour on the ten frame. Then place 8 counters of another colour. Use the counters to help you solve the number sentences.

6 + 8 = _____ so 8 + 6 = _____ so _____ – 8 = 6 so _____ – 6 = 8

2 Use the ten frames and counters to help you solve the number sentences. Then record all the possible number sentences.

11 + 6 = _____ so _____ so _____ so _____

9 + 7 = _____ so _____ so _____ so _____

5 + 14 = _____ so _____ so _____ so _____

7 + 6 = _____ so _____ so _____ so _____

9 + 2 = _____ so _____ so _____ so _____

8 + 4 = _____ so _____ so _____ so _____

6 + 5 = _____ so _____ so _____ so _____

7 + 8 = _____ so _____ so _____ so _____

5 + 8 = _____ so _____ so _____ so _____

Unit 16 **Addition and Subtraction** (TRB pp. 82–85)
Addition and subtraction MA1-5NA uses a range of strategies and informal recording methods for addition and subtraction involving one- and two-digit numbers

Missing Numbers

You will need: a partner, a copy of BLM 1 'Make a Number Line'

1 With a partner, make a number line from 0 to 20 using BLM 1.

Use your number line to help you solve the following problems.

13 + ☐ = 19   8 + ☐ = 12   7 + ☐ = 15
19 − ☐ = 13   12 − ☐ = 8   15 − ☐ = 7

9 + ☐ = 15   5 + ☐ = 13   11 + ☐ = 18
15 − ☐ = 9   13 − ☐ = 5   18 − ☐ = 11

14 + ☐ = 17   6 + ☐ = 13   12 + ☐ = 17
17 − ☐ = 14   13 − ☐ = 6   17 − ☐ = 12

7 + ☐ = 16   13 + ☐ = 17   9 + ☐ = 17
16 − ☐ = 7   17 − ☐ = 13   17 − ☐ = 9

11 + ☐ = 16   4 + ☐ = 12   12 + ☐ = 20
16 − ☐ = 11   12 − ☐ = 4   20 − ☐ = 12

2 Solve the following problem, and show how you worked it out.
Eighteen people were on a bus. Some people
got off, which left 13 people on the bus.
How many people got off the bus?

Unit 16 **Addition and Subtraction** (TRB pp. 82–85)
Addition and subtraction MA1-5NA uses a range of strategies and informal recording methods for addition and subtraction involving one- and two-digit numbers

65

Four Fact Family

You will need: two 10-sided dice

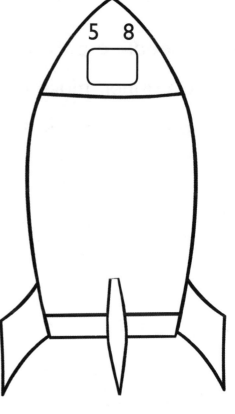

1 a Look at the top of the rocket. Add the 2 numbers together and write the total in the box.

b Use your 3 numbers to make **four fact family number sentences**. Write your number sentences inside the rocket.

2 Complete the rockets below by rolling two 10-sided dice. Write the numbers and their total in the top of each rocket. Then write the **four fact family number sentences** inside each rocket.

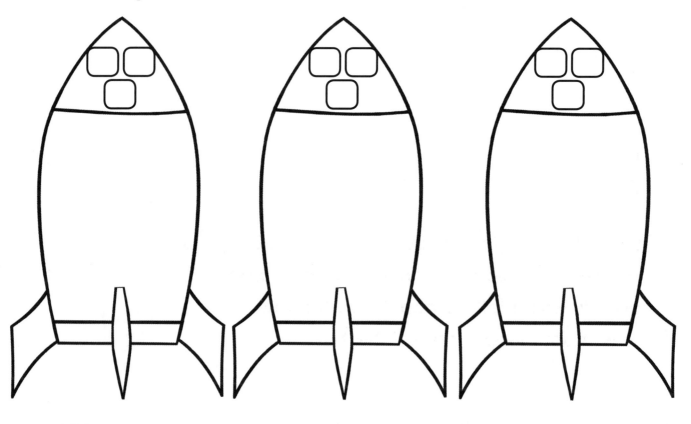

 Addition and Subtraction (TRB pp. 82–85)
Addition and subtraction MA1-5NA uses a range of strategies and informal recording methods for addition and subtraction involving one- and two-digit numbers

DATE:

STUDENT ASSESSMENT

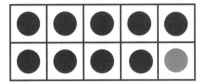 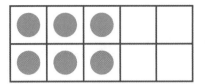

1 Look at the ten frames and fill in the number sentences.

9 + 7 = ⬚   so   ⬚ + ⬚ = ⬚

and ⬚ – ⬚ = ⬚   so   ⬚ – ⬚ = ⬚

2 a Show 11 + 4 on the ten frames below.

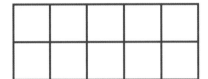 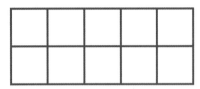

b Use your answer to Question 2a to help you complete the number sentences.

⬚ + ⬚ = ⬚   ⬚ + ⬚ = ⬚

⬚ – ⬚ = ⬚   ⬚ – ⬚ = ⬚

3 Use the number line to help you solve the problems below.

18 – ⬚ = 13   so   18 – ⬚ = 5

and ⬚ + ⬚ = 18   so   ⬚ + ⬚ = ⬚

4 a Add the numbers shown on the

     10-sided dice. _____

b Now use the 3 numbers from Question 4a to write four fact family number sentences.

Unit 16   **Addition and Subtraction** (TRB pp. 82–85)
Addition and subtraction MA1-5NA uses a range of strategies and informal recording methods for addition and subtraction involving one- and two-digit numbers

67

Money! Money!

You will need: scissors, a copy of BLM 36 'Coins' and BLM 37 'Notes', glue

1 Cut out the coins and notes from BLM 36 and BLM 37. Paste them under the correct amounts.

$100	50c	20c
ten dollars	five cents	$1
$2	$50	five dollars
twenty cents	twenty dollars	10c

2 Design a new $200 note.

front

back

Money (TRB pp. 86–89)
Whole numbers (money) MA1-4NA applies place value, informally, to count, order, read and represent two- and three-digit numbers

How Might it Look?

You will need: scissors, a copy of BLM 36 'Coins' and BLM 37 'Notes', glue

I Look at each amount of money. Use coins and notes from BLM 36 and BLM 37 to make the same amount another way. Paste your coins and notes in the boxes.

a

b

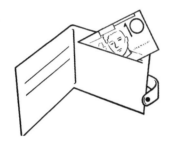

c

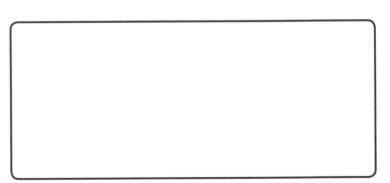

2 If you had , what items could you buy? List the items below.

_____

_____

Unit 17 **Money** (TRB pp. 86–89)
Whole numbers (money) MA1-4NA applies place value, informally, to count, order, read and represent two- and three-digit numbers

69

Coins! Coins! Coins!

You will need: scissors, a copy of BLM 36 'Coins', glue

1 How much money is in each piggy bank?

_____   _____

2 What coins would you use to pay for each stamp?
Cut out coins from BLM 36 and paste them below each stamp.

70c
AUSTRALIA

45c
AUSTRALIA

3 Use 3 different coins from BLM 36 and show:

a the **highest** amount you can make	**b** the **lowest** amount you can make.
How much altogether? _____	How much altogether? _____

Unit 17

STUDENT ASSESSMENT

You will need: scissors, 3 copies of BLM 36 'Coins', a copy of BLM 37 'Notes', glue

1 Cut out the coins and notes from BLMs 36 and 37. Paste them in the right places, under the amounts below.

$2	20c	twenty dollars

$50	fifty cents	$5

2 Write how much each group of coins is worth.

_____   _____

3 a Cut out the coins from BLM 36. Paste down the coins that you would use to pay for the strawberries.

65c

b Use the coins from BLM 36 to show another way that you could make 65c.

Unit 17 **Money** (TRB pp. 86–89)
Whole numbers (money) MA1-4NA applies place value, informally, to count, order, read and represent two- and three-digit numbers

71

Numbers and Shapes

DATE:

You will need: pattern blocks, a sheet of paper, coloured pencils

1 Continue the number patterns.

2, 4, 6, 2, 4, 6, 2, 4, 6, _____, _____, _____, _____, _____, _____, _____

3, 1, 1, 3, 1, 1, 3, 1, 1, _____, _____, _____, _____, _____, _____, _____

2, 4, 1, 4, 2, 4, 1, 4, 2, 4, _____, _____, _____, _____, _____, _____, _____

2 Continue the shape patterns.

3 On another sheet of paper, trace over some pattern blocks to match the number pattern.

2   1   3   2   1   3

4 On another sheet of paper, create your own number pattern. Then trace over some pattern blocks to make a shape pattern to match.

Skip Counting to Make Patterns

You will need: a copy of BLM 1 'Make a Number Line', scissors, glue

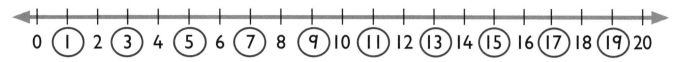

1 a Write the numbers that are circled on the number line.

_____

b Write what you think the next 4 numbers will be.

_____

c Explain how you worked out your answer to Question 1b.

_____

2 Make a number line using BLM 1. Start your number line at 0.

3 Use your number line to help you complete the patterns.

a 3, 8, 13, 18, _____, _____, _____, 38, _____, 48

b 6, 16, 26, _____, _____, 56, 66, _____, _____

c 19, 21, _____, _____, 27, 29, _____, _____, _____, 37

d 38, 36, _____, 32, 30, _____, _____, _____, 22, _____

e 14, 19, 24, _____, _____, _____, _____, 49, _____

4 Use your number line to make the following patterns:

a a pattern skip counting by 5s

_____

b a pattern skip counting by 2s

_____

c a pattern skip counting by 10s

_____

Unit **18** **Number Patterns** (TRB pp. 90–93)
Patterns and algebra MA1-8NA creates, represents and continues a variety of patterns with numbers and objects

73

Adding to Make Patterns

1 Add the numbers to complete the pattern below.

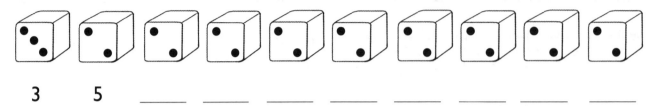

3   5   ___ ___ ___ ___ ___ ___ ___ ___

Explain what you notice about the pattern.

_____

2 Trace over the remaining matchstick in the pattern. Continue to add on sticks to complete the number pattern.

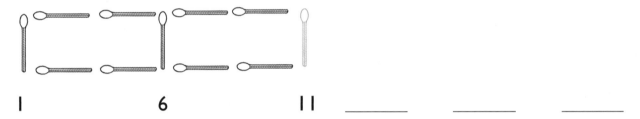

1   6   11   _____   _____   _____

3 a Complete the pattern.

8   18   28   ___   ___   ___   ___

b What do you notice about the pattern?

_____

4 a Complete the pattern below.

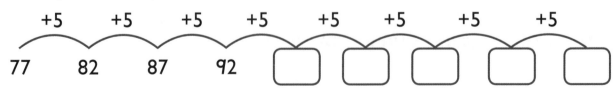

+5   +5   +5   +5   +5   +5   +5   +5

77   82   87   92   ◯   ◯   ◯   ◯   ◯

b Explain how you know your pattern is correct.

_____

STUDENT ASSESSMENT

1 Write the number pattern that matches this pattern.

_____

2 Draw a pattern to match this number pattern.

3   I   3   I   3   I   3   I

3 Use the number line to show a pattern that skip counts by 2.

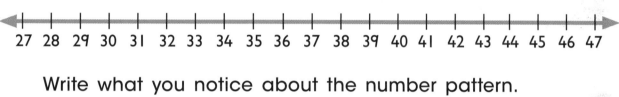

27 28 29 30 31 32 33 34 35 36 37 38 39 40 41 42 43 44 45 46 47

Write what you notice about the number pattern.

_____

4 Complete the pattern to match the dice.

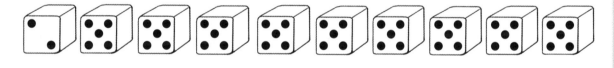

2   7   12   _____ _____ _____ _____ _____ _____ _____

Explain what you notice about the pattern.

_____

5 Write the missing numbers in the number patterns.

23, 33, 43, _____, _____, 73, _____, _____, 103, _____

45, 47, 49, _____, 53, _____, 57, _____, _____, 63

64, 69, 74, _____, _____, _____, _____, _____, 104, _____

Unit
18

Number Patterns (TRB pp. 90–93)
Patterns and algebra MA1-8NA creates, represents and continues a variety of patterns with numbers and objects

75

Animals on the Farm

1 A farmer had 16 sheep on her farm. She kept the sheep in 2 paddocks. Draw 2 of the ways that she could have placed her sheep in the paddocks.

a

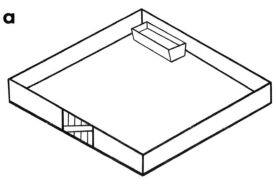

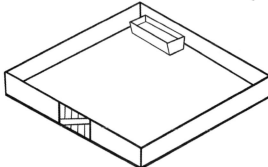

b

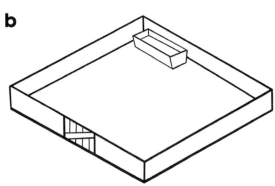

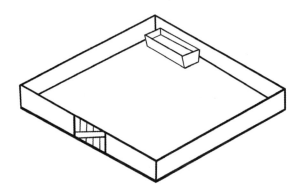

2 Write a number sentence below each of your combinations in Question 1.

3 Write number sentences for all the other number combinations you could have drawn.

Unit 19 More Number Patterns (TRB pp. 94–97)
Patterns and algebra MA1-8NA creates, represents and continues a variety of patterns with numbers and objects

Word Problems

1 Read the word problems in the table. In the second column,
write the number sentence for each problem and solve it.
In the third column, draw a picture of the problem.

Word problem	Number sentence and solution	Picture
A bird laid 13 eggs. 7 eggs hatched. How many more needed to hatch?		
Your class is going on a bus. 5 students get on the bus. How many more students need to get on the bus?		
7 beetles were hiding under some leaves in a jar. 12 more beetles could be seen. How many beetles were there altogether?		
Angus had a packet of 24 pencils. He lost 7 pencils. How many did he have left?		

2 How did you work out the number sentences?

_____

Unit 19 **More Number Patterns** (TRB pp. 94–97)
Patterns and algebra MA1-8NA creates, represents and continues a variety of patterns with numbers and objects

77

Think Board

You will need: a card from a copy of BLM 43 'Number Sentences'

Write the number sentence from the card, and complete the rest of the think board.

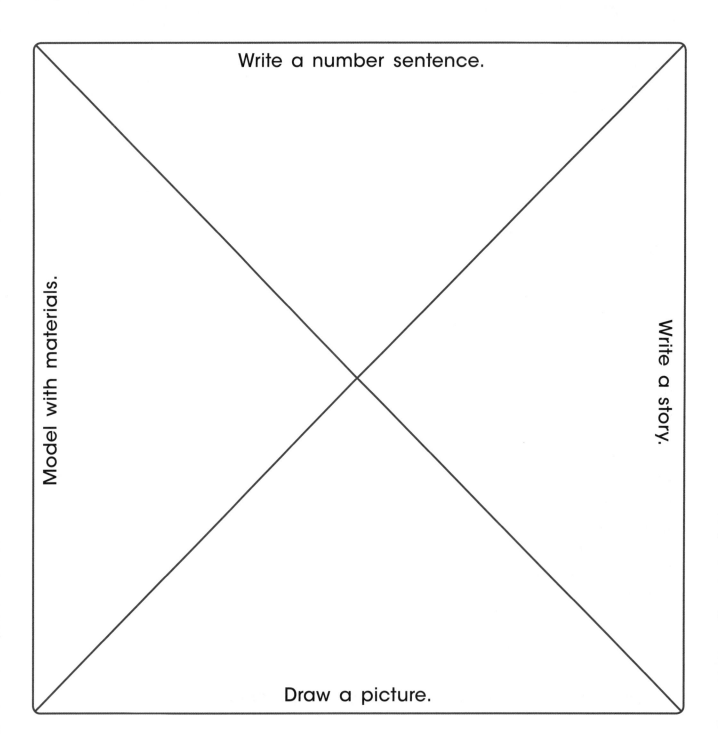

Write a number sentence.

Model with materials.

Write a story.

Draw a picture.

DATE:

STUDENT ASSESSMENT

1 a Ram had 6 counters, some were red and some were blue. Draw how his counters may have looked. Write the number sentence to match.

b Have you shown all the combinations that Ram could have had? _____ How do you know?

_____

2 Jo picked 8 apples from one tree and 9 apples from another tree. How many apples did she pick altogether? Write a number sentence for this problem and solve it.

_____

3 Write a word problem for each of the following.

a $16 - 9 = 7$

_____

_____

b $8 + 16 = 24$

_____

_____

How Much?

You will need: a plastic bottle; a cup for measuring; rice, sand or water; a selection of containers (labelled A, B, C, etc.)

> Before you begin this page, choose rice, sand or water to measure.
>
> I will use _____

1 Look at the bottle and the cup. How many cups do you think it will take to fill the bottle?

My guess is _____ cups.

2 Measure how many cups of your chosen item the bottle holds, and fill in your answer below.

= _____ cups

3 Choose 4 other containers that you think will hold the **same amount** as the bottle.

a Find out how many cups of your chosen item each container holds, then fill in the table.

Name of container	How many cups it holds

b Write what you found out.

_____

c Explain how you chose your containers.

_____

 Capacity (TRB pp. 98–101)
Volume and capacity MA1-11MG measures, records, compares and estimates volumes and capacities using uniform informal units

How Many Scoops?

You will need: a selection of containers
(labelled A, B, C, etc.); rice, sand or water; measuring scoop

> Before you begin this page, choose rice, sand or water to measure.
>
> I will use _____

1 Choose 4 containers. Estimate and then measure how many
scoops of your chosen item each container will hold, and fill in
the table.

Name of container	Estimate	Actual measure

2 How did you estimate how much each container would hold?

_____

3 a Which container holds the **most**? _____

b Which container holds the **least**? _____

4 a How much **more** than the **smallest** container does the
largest container hold? _____

b Show how you worked out your answer to Question 4a.

Unit 20 **Capacity** (TRB pp. 98–101)
Volume and capacity MA1-11MG measures, records, compares and estimates volumes and capacities using
uniform informal units

81

Measuring Jugs

You will need: a selection of measuring containers; sand, rice or water

1 Record how much liquid is in each container.

a

_____

b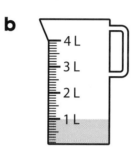

_____

c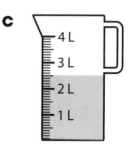

_____

d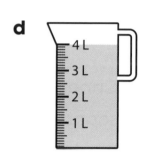

_____

e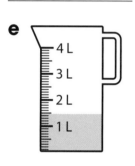

_____

f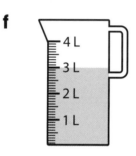

_____

g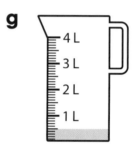

_____

h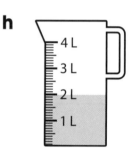

_____

2 Look at the amount in the jug, then complete the questions.

a Find a container in the room that holds:

	Name or picture of container
more than the jug	
less than the jug	
the same amount as the jug	

b How did you choose your containers?

_____

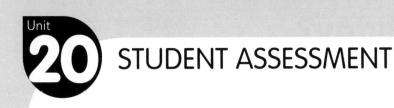

STUDENT ASSESSMENT

You will need: a bottle; a scoop; a selection of containers (labelled A, B, C, etc.); rice, sand or water; a 1-litre measuring container

Before you begin this page, choose rice, sand or water to measure.

I will use _____

1 Look at the bottle. Estimate and then measure how many **scoops** of your chosen item it will take to fill the bottle, and complete the table.

Bottle	Estimate	Actual measure
	_____ scoops	_____ scoops

2 a Choose another container, and circle the statement that you think is correct.

| The container holds **more than** the bottle | The container holds **the same amount as** the bottle | The container holds **less than** the bottle |

b Use the scoop to check if you circled the correct statement. Were you correct? _____

3 Look at the bottle, your second container and 2 more containers. Draw or write the name of each container where you think it belongs in the table.

Less than 1 litre	1 litre	More than 1 litre

Capacity (TRB pp. 98–101)
Volume and capacity MA1-11MG measures, records, compares and estimates volumes and capacities using uniform informal units

83

When Is Your Birthday?

You will need: a calendar

1 Make a calendar for your birthday month.
First, write the name of the month.

2 Next, write the days of the week in the first row.

3 Look at a calendar. Work out what day of the week the month begins on, and how many days are in the month.

4 Fill in the numbers.

5 Circle the date of your birthday.

6 Write any other special days that occur in your birthday month.

_____

7 Which day is your birthday on? _____

8 What is the date of the second Wednesday of the month? _____

9 How many Thursdays are there in the month? _____

The month of _____

Monday						

More About Time (TRB pp. 102–105)
Time MA1-13MG describes, compares and orders durations of events, and reads half- and quarter-hour time

Days in the Month

Complete the calendar. Think of the chant '30 Days has September' to help you remember the number of days in each month.

Year _____

January						
M	T	W	Th	F	S	Sun

February						
M	T	W	Th	F	S	Sun

March						
M	T	W	Th	F	S	Sun

April						
M	T	W	Th	F	S	Sun

May						
M	T	W	Th	F	S	Sun

June						
M	T	W	Th	F	S	Sun

July						
M	T	W	Th	F	S	Sun

August						
M	T	W	Th	F	S	Sun

September						
M	T	W	Th	F	S	Sun

October						
M	T	W	Th	F	S	Sun

November						
M	T	W	Th	F	S	Sun

December						
M	T	W	Th	F	S	Sun

Unit 21 **More About Time** (TRB pp. 102–105)
Time MA1-13MG describes, compares and orders durations of events, and reads half- and quarter-hour time

85

Seasons

1 Fill in the missing months. Remember to keep them in the correct order.

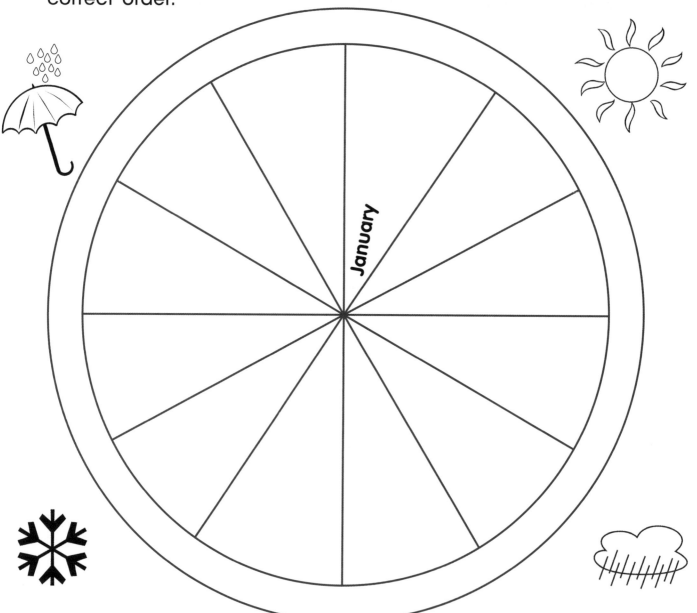

January

2 Divide the outer ring into the seasons, and label each season.

3 Write about your favourite season; for example, what it is like and what you like to do. _____

_____

_____

STUDENT ASSESSMENT

DATE:

1 Write the months of the year that begin with the letter 'J'.

_____

2 Which month comes before November? _____

3 Which month comes after March? _____

4 Which month is your birthday in? _____

5 Which is the last month of the school year? _____

6 How many days are in each of the following months?

   a March _____   **b** May _____

   c August _____   **d** September _____

   e November _____

7 **a** Which season is it now? _____

   b Which months are in that season? _____

   _____

   c Which season comes next? _____

8 Draw a picture of your favourite season and add a label.

Unit 21 **More About Time** (TRB pp. 102–105)
Time MA1-13MG describes, compares and orders durations of events, and reads half- and quarter-hour time

87

Pet Photo Day

You will need: a copy of BLM 48 'Pets', scissors, glue

BLM 48 shows the pet photos that students have brought to school for the 'Best Pet Competition'.

I Cut out the photos and organise them so that it is easy to see the different types of pets, and how many there are of each type. Paste the photos below.

2 Who do you think should win the 'Best Pet Competition'?

Unit 22 **Data** (TRB pp. 106–109)
Data MA1-17SP gathers and organises data, displays data in lists, tables and picture graphs, and interprets the results

How Many Cubes?

You will need: interlocking cubes

1 Grab as many cubes as you can in your left hand. Count how many cubes you picked up and fill in the table below.

Name	Me									
How many cubes?										

2 Ask 9 other students to pick up a handful of cubes. In the table, record their name and the number of cubes they picked up.

3 Make a graph to show how many cubes each person grabbed.

Title: _____

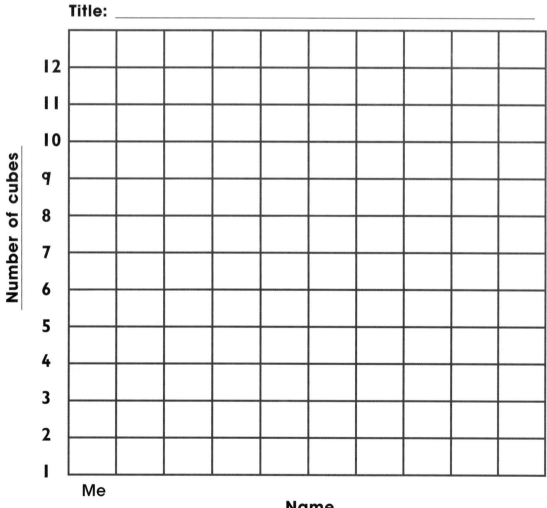

Number of cubes

12
11
10
9
8
7
6
5
4
3
2
1

Me

Name

Unit **22** **Data** (TRB pp. 106–109)
Data MA1-17SP gathers and organises data, displays data in lists, tables and picture graphs, and interprets the results

89

Going to School

1 Total the tally marks in the table.

Category	Walk	Bus	Bike	Car
Tally	✝✝✝✝ ||||	||||	|||	✝✝✝✝ |||
Total				

2 Make a graph of the information in the table.

How Children Come to School

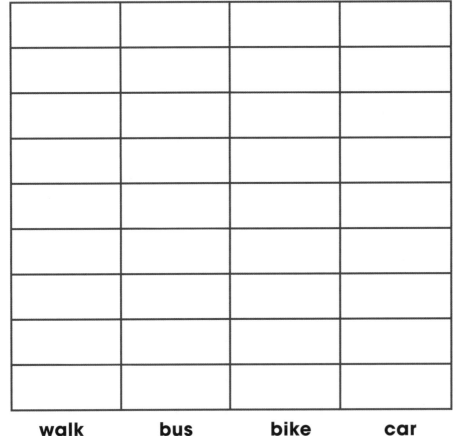

walk   bus   bike   car

3 Write 3 things about your graph.

_____

_____

_____

Data (TRB pp. 106–109)
Data MA1-17SP gathers and organises data, displays data in lists, tables and picture graphs, and interprets the results

STUDENT ASSESSMENT

I Write the total for each colour in the table.

Category	Tally	Total
Red	◎◎◎ I	
Green	◎◎◎	
Yellow	|||	
Blue	◎◎◎ ||||	

2 Make a graph of the information in the table.

Title: _____

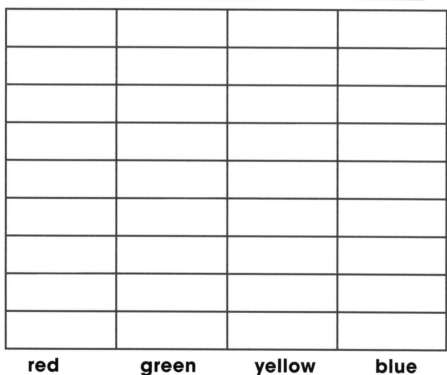

red   green   yellow   blue

_____

3 Think of a question that could have been asked in order to get the information shown in the table.

_____

4 What does the graph show?

_____

Unit
22
Data (TRB pp. 106–109)
Data MA1-17SP gathers and organises data, displays data in lists, tables and picture graphs, and interprets the results

91

Lots of Animals

DATE:

1 Draw 6  in each bowl.

How many fish are
there altogether? _____

2 Draw 3 ◯ in each nest.

How many eggs are
there altogether? _____

3 Draw 2 🐝 on each flower.

How many bees are
there altogether? _____

4 Draw 7 🐦 in each tree.

How many birds are
there altogether? _____

5 Draw 4 🐱 in each basket.

How many kittens are
there altogether? _____

6 Draw a picture for
the number sentence.

5 + 5 + 5

How many are there altogether? _____

Unit **23** **Multiplication** (TRB pp. 110–113)
Multiplication and division MA1-6NA uses a range of mental strategies and concrete materials for multiplication and division

Cups and Counters

You will need: a partner, a dice, paper cups, counters

1 Work with a partner. Roll the dice and count out that many cups.

2 Roll the dice again and put that many counters into each cup. Complete the statement.

Cups   *Counters*

[   ] multiplied by [   ] makes [   ]

3 Repeat steps 1 and 2.

Cups   *Counters*

[   ] multiplied by [   ] makes [   ]

[   ] multiplied by [   ] makes [   ]

[   ] multiplied by [   ] makes [   ]

[   ] multiplied by [   ] makes [   ]

4 If I had 3 bags of marbles with 7 marbles in each bag, how many marbles did I have altogether? _____

Show how you solved the problem.

Unit 23 **Multiplication** (TRB pp. 110–113)
Multiplication and division MA1-6NA uses a range of mental strategies and concrete materials for multiplication and division

93

Pictures and Problems

1 Write a number sentence for each group of pictures.

Solve each number sentence.

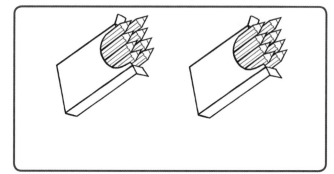

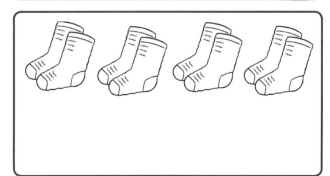

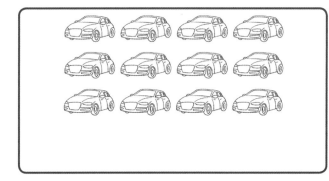

2 Draw a picture for each number sentence.

Solve each number sentence.

5 × 3 =	1 × 7 =	4 × 4 =
8 × 2 =	2 × 4 =	3 × 6 =

Unit 23 Multiplication (TRB pp. 110–113)
Multiplication and division MA1-6NA uses a range of mental strategies and concrete materials for multiplication and division

DATE:

STUDENT ASSESSMENT

You will need: counters

I Draw 3 in each web.

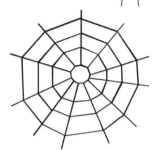 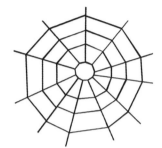 

How many spiders are there altogether? _____

2 Draw 2 in each bowl.

How many apples are there altogether? _____

3 **a** Draw a picture to show **6 multiplied by 3**.

b Write the answer to Question 3a. _____

4 **a** Use counters to solve 5 × 4. _____

b Draw how you solved the problem.

Unit
23
Multiplication (TRB pp. 110–113)
Multiplication and division MA1-6NA uses a range of mental strategies and concrete materials for multiplication and division

95

Arrays

1 Write a number sentence for each array.

Solve each number sentence.

_____ _____ _____ _____

_____ _____ _____ _____

2 Draw an array for each number sentence.

Solve each number sentence.

4 × 4 =	9 × 1 =
2 × 10 =	6 × 4 =

3 Write a number sentence

for the box of pears.

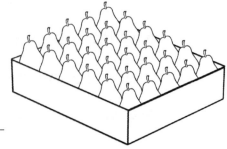

_____

Work out how many pears are in the box. _____

Comparing Arrays

You will need: a copy of BLM 53 'Dots', scissors, glue

1 Draw an array for:

7 × 3   3 × 7

The answer is _____

2 Cut out an array of dots from BLM 53. Paste the array below.

Write a matching number sentence and solve it.

_____

3 Use BLM 53, and cut out the matching arrays for the following number sentences. Paste them below the number sentences.

a 2 × 5   5 × 2

The answer is _____

b 6 × 4   4 × 6

The answer is _____

4 What did you notice about arrays?

_____

Unit 24 **More About Multiplication** (TRB pp. 114–117)
Multiplication and division MA1-6NA uses a range of mental strategies and concrete materials for multiplication and division

97

Multiply by 2

You will need: a 10-sided dice (0–9), a partner

Take turns to roll the dice. Multiply the number that you roll by 2, and colour that number on your caterpillar. The first person to colour all the numbers on their caterpillar wins.

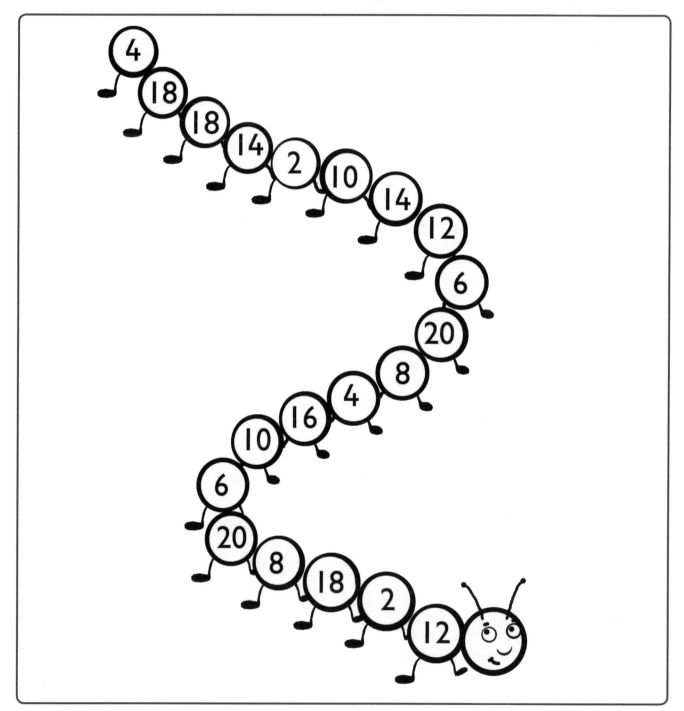

1 Write a number sentence for each array. Solve each number sentence.

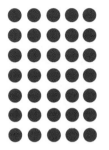  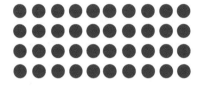

_____   _____   _____

2 Draw an array for each number sentence. Solve each number sentence.

3 × 4 =	5 × 6 =	7 × 3 =

3 4 × 7 = 28   so   7 × 4 = _____

4 a Solve the number sentences, without drawing arrays.

8 × 2 = _____   6 × 2 = _____   9 × 2 = _____

b How did you work out the answers to the number sentences?

_____

Unit
24
More About Multiplication (TRB pp.114–117)
Multiplication and division MA1-6NA uses a range of mental strategies and concrete materials for multiplication and division

99

Largest Footprint

You will need: a partner, paper, scissors

1 On some paper, have your partner trace around your foot (leave your socks on). Cut out your footprint and paste it in the box.

2 Compare your footprint with your partner's footprint. Whose footprint has the largest area? _____

3 Find something in the room that has a **greater area** than your footprint and something that has a **smaller area**.

   a On paper, draw the item that has a **greater area** than your footprint.

   b On paper, draw the item that has a **smaller area** than your footprint.

Footprints

You will need: a partner, paper, scissors

1 On some paper, have your partner trace around your foot (leave your socks on). Cut out your footprint. You will use your footprint to work out the area of some objects.

2 a Estimate how many footprints you will need to cover the items in the table. Add your estimates to the table.

b Now measure how many footprints it takes to cover the items in the table. Add your measurements to the table.

	Estimate	Measurement
book cover		
top of table		
seat of chair		
item of your choice: _____		

3 Find something in the classroom that has the same area as your footprint. On paper, draw and label it.

4 Is it good to measure with footprints? _____

   Why? _____

Unit **25** **Area** (TRB pp. 118–121)
Area MA1-10MG measures, records, compares and estimates areas using uniform informal units

101

Wrapping Paper

1 Jack wanted to put some shapes in size order, so he traced them onto wrapping paper.

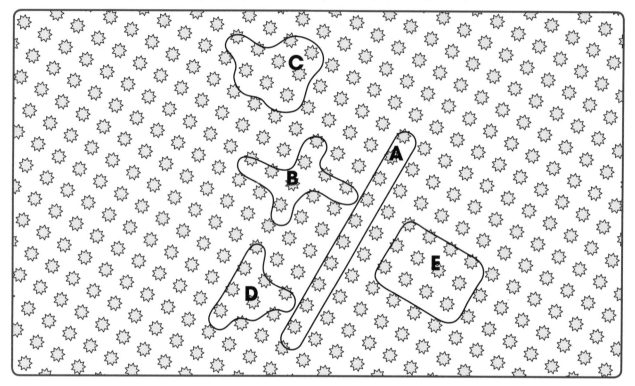

a Order the shapes from the one with the **largest area** to the one with the **smallest area**.

_____

b How did you work out the answer to Question 1a?

_____

_____

2 a Which combination of shapes would have approximately the same area as your handprint?

_____

b Explain how you worked out the answer to Question 2a.

_____

_____

Area (TRB pp. 118–121)
Area MA1-10MG measures, records, compares and estimates areas using uniform informal units

STUDENT ASSESSMENT

You will need: paper, scissors, a lid from  a large tub, a packet of playing cards

1 On paper, trace around your hand (keep your fingers and thumb together) and your foot (leave your socks on). Cut out your handprint and footprint.

2 a Which has the greater area — your footprint or **2** handprints? _____

    b Explain how you worked out which has the greater area. _____

_____

3 a How many handprints is the area of the lid from the tub? _____

    b Draw how you worked out the answer to Question 3a.

4 a Use playing cards to work out the area of the top of your table. _____

    b Draw how you worked out the answer to Question 4a.

Unit
25
Area (TRB pp. 118–121)
Area MA1-10MG measures, records, compares and estimates areas using uniform informal units

103

Name the 3D Objects

1 Draw a line from each object to its name.

    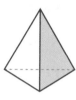

> triangular prism   cylinder   cone   cube   rectangular prism

2 Which 3D object is this ? _____

3 This is the shape of a ball.

Why is it a good shape for a ball?

_____

4 Why is this ▯ a good shape for a can?

_____

5 Look at the arrangements of blocks. Circle the arrangement that is most likely to fall down.

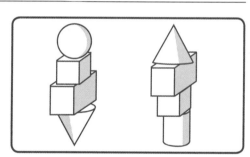

6 a Which 3D objects would you use to make a building that is stable? Name and draw the objects.

b Explain why you chose those objects.

_____

3D Objects (TRB pp. 122–125)
Three-dimensional space MA1-14MG sorts, describes, represents and recognises familiar three-dimensional objects, including cones, cubes, cylinders, spheres and prisms

Features of 3D Objects

You will need: a set of 3D objects made from a copy of BLM 56 '3D Objects I', BLM 57 '3D Objects 2' and BLM 58 '3D Objects 3'

I a In a small group, look at the tetrahedron and the cube. In the table, write everything you know about these 3D objects. If the objects share any features, write these in the middle column.

Features of a **tetrahedron** (include faces, corners and edges)	Similar features of both objects	Features of a **cube** (include faces, corners and edges)

b What could you use a tetrahedron for?

_____

c What could you use a cube for?

_____

2 Choose 2 other objects to compare.

Features of a _____ (include faces, corners and edges)	Similar features of both objects	Features of a _____ (include faces, corners and edges)

Unit 26 **3D Objects** (TRB pp. 122–125)
Three-dimensional space MA1-14MG sorts, describes, represents and recognises familiar three-dimensional objects, including cones, cubes, cylinders, spheres and prisms

105

Look and Sort

You will need: a set of 3D objects made from a copy of BLM 56 '3D Objects 1', BLM 57 '3D Objects 2' and BLM 58 '3D Objects 3'

1 Sort the objects, and write which ones belong to each group below.

Objects with a **curved** surface	Objects with a **flat** surface

2 Sort the objects again, and write which ones belong to each group below.

Objects **with** corners	Objects **without** corners

3 Here are some shadows of 3D objects. What 3D objects could they be?

_____

_____

_____

_____

3D Objects (TRB pp. 122–125)
Three-dimensional space MA1-14MG sorts, describes, represents and recognises familiar three-dimensional objects, including cones, cubes, cylinders, spheres and prisms

DATE:

STUDENT ASSESSMENT

You will need: a set of 3D objects made from a copy of BLM 56 '3D Objects I', BLM 57 '3D Objects 2' and BLM 58 '3D Objects 3'

I Name the 3D objects.

_____ _____ _____ _____ _____

2 Complete the table.

	Tetrahedron	Cube	Triangular prism
number of faces			
number of edges			
number of corners			

3 A 3D object that has a curved surface is inside the bag. What could it be?

_____

4 A 3D object that has 8 corners is inside the bag. What could it be?

_____

Unit
26
3D Objects (TRB pp. 122–125)
Three-dimensional space MA1-14MG sorts, describes, represents and recognises familiar three-dimensional objects, including cones, cubes, cylinders, spheres and prisms

107

Sharing

1 There are 15 seeds in a packet.
   The seeds need to be shared
   equally between 3 pots.
   Draw the seeds in each pot.

2 There are 20 fish in a bag.
   The fish have to be shared
   equally into 5 bowls.
   Draw the fish in the bowls.

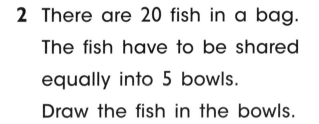

3 There are 16 biscuits in a packet.
   The biscuits have to be shared
   equally onto 4 plates.
   Draw the biscuits on the plates.

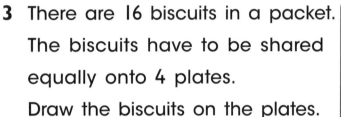

4 There are 12 pencils in a bundle.
   They need to be shared
   equally into 2 tins.
   Draw the pencils in the tins.

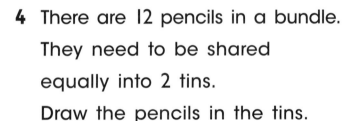

5 Show how you would share
   some stickers with your friends.

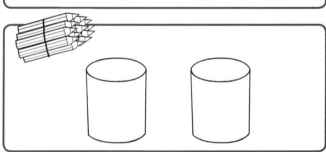

6 Write your own sharing story.
   Add some pictures.

   _____

   _____

   _____

   _____

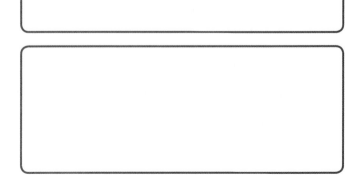

Division (TRB pp. 126–129)
Multiplication and division MA1-6NA uses a range of mental strategies and concrete materials for multiplication and division

Share to Solve

You will need: counters

1 Use counters to help you solve the problems.

   a 12 shared equally between 6 is _____

   b 22 shared equally between 2 is _____

   c 24 shared equally between 3 is _____

   d 12 shared equally between 4 is _____

   e 28 shared equally between 4 is _____

   f 20 shared equally between 10 is _____

   g 20 shared equally between 5 is _____

   h 24 shared equally between 2 is _____

   i 21 shared equally between 7 is _____

   j 28 shared equally between 7 is _____

   k 24 shared equally between 6 is _____

   l 20 shared equally between 4 is _____

   m 21 shared equally between 3 is _____

   n 20 shared equally between 2 is _____

2 There were 24 cards in a pack. The cards were dealt equally to 8 children. How many cards did each child get? _____

3 16 horses arrived at a farm. They were placed equally into 4 paddocks. How many horses were in each paddock? _____

Unit 27 **Division** (TRB pp. 126–129)
Multiplication and division MA1-6NA uses a range of mental strategies and concrete materials for multiplication and division

109

Number Mat Game

You will need: a partner, cards made from a copy of BLM 62 'Division Cards', counters, coloured pencils

To play this game, each player needs to choose a different coloured pencil.

1 Place the Division Cards face down in a pile.

2 In turn, select a card and work out the answer.
(You can use counters to help you.)

3 Colour the answer to the Division Card on the number mat.

4 The first person to colour 4 numbers in a row wins.

5 Now play the game on your partner's number mat.

3	4	5	2
6	1	4	5
2	4	2	3
3	6	5	1

Division (TRB pp. 126–129)
Multiplication and division MA1-6NA uses a range of mental strategies and concrete materials for multiplication and division

STUDENT ASSESSMENT

You will need: counters

1 a Draw how you would share 12  equally between 2 bowls.

b Write the number sentence for this problem.

2 a Draw how you would share 9  equally between 3 bags.

b Write the number sentence for this problem.

3 Use counters to work out some other division problems that have the **same answer** as Question 1.
Write the number sentence for each problem.

Unit
27
Division (TRB pp. 126–129)
Multiplication and division MA1-6NA uses a range of mental strategies and concrete materials for multiplication and division

111

Draw and Group

1 Here are 18 crayons.

How many packets will there be, if there are 6 crayons in each packet?

2 Draw 12 carrots.

How many bunches will there be, if there are 3 carrots in a bunch?

3 Draw 20 playing cards.

How many hands of cards will there be, if there are 5 cards in each hand?

4 Draw 12 eggs.

How many boxes of eggs will there be, if there are 6 eggs in a box?

5 Draw 16 straws.

How many squares will you make if you join the straws together?

How Many Groups?

You will need: counters

1 Write the number sentence and the answer for each
of the problems. Use counters to help you.

a 36 How many groups of 6? _____

b 22 How many groups of 11? _____

c 16 How many groups of 2? _____

d 28 How many groups of 4? _____

e 25 How many groups of 5? _____

f 15 How many groups of 3? _____

g 36 How many groups of 9? _____

h 27 How many groups of 3? _____

i 39 How many groups of 3? _____

j 30 How many groups of 10? _____

2 When Bo solved her problem, she had 4 groups.

Write 3 number sentences for problems Bo could have solved.

_____

_____

_____

Unit 28 **More About Division** (TRB pp. 130–133)
Multiplication and division MA1-6NA uses a range of mental strategies and concrete materials for multiplication and division

113

Share the Chocolate

1 Look at the block of chocolate.

Draw how many different ways you could share the chocolate fairly.

2 Look at the larger block of chocolate.

Draw how many different ways you could share the chocolate fairly.

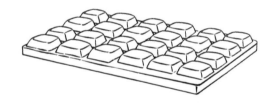

STUDENT ASSESSMENT

1 a Draw 15 flowers.

b In the space below, show how many bunches of 5 flowers you can make.

c Write the number sentence for the problem and your answer.

_____

d What other **even** bunches could you make with your 15 flowers? Draw your solutions.

2 There are 14 tiny teddy biscuits in a packet. Draw them.

a How many groups of 2 can you make? _____

b Write the number sentence for the problem and your answer. _____

c What other ways could the tiny teddy biscuits be grouped? Draw your solutions.

Unit
28
More About Division (TRB pp. 130–133)
Multiplication and division MA1-6NA uses a range of mental strategies and concrete materials for multiplication and division

115

Cross Off 6

You will need: a dice, a partner, coloured pencils

1 a If you roll a dice, what number could it land on? _____

  b With your partner, take turns to roll the dice and cross off the matching number on the list below. If the number is already crossed off, you will have to miss a turn. The first person to cross off all **6** numbers wins.

| 1 | 2 | 3 | 4 | 5 | 6 |

  c What was the last number you crossed off? _____

2 a If you played the game again, do you think the last number you crossed off would be the same this time? _____
Why? Why not? _____

_____

  b Play the game again.

| 1 | 2 | 3 | 4 | 5 | 6 |

  c Was the game the same as last time? Explain.

_____

_____

_____

Chance (TRB pp. 134–137)
Chance MA1-18SP recognises and describes the element of chance in everyday events

In the Classroom

1 Look at the picture that shows things happening in the classroom. Cross out the things that are **impossible**.

2 Draw something that you are **certain** you would see outside your classroom window.

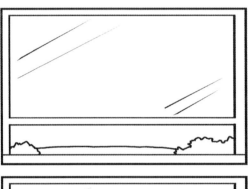

3 Draw something that would be **impossible** to see outside your classroom window.

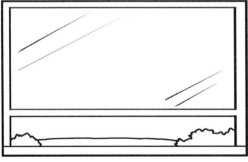

4 Draw something that **might happen** outside your classroom window.

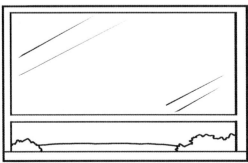

Unit 29 **Chance** (TRB pp. 134–137)
Chance MA1-18SP recognises and describes the element of chance in everyday events

117

What Do You Think?

You will need: a copy of BLM 65 'What's the Chance?', scissors, glue

1 Cut out the statements from BLM 65 and paste each under a category heading.

Certain

Likely

Unlikely

Impossible

2 Do you think your list will be the same as that of the other students in the class? Why? Why not?

_____

_____

Chance (TRB pp. 134–137)
Chance MA1-18SP recognises and describes the element of chance in everyday events

DATE:

STUDENT ASSESSMENT

1 When you toss a coin do you know if it will land on heads or tails? Explain.

_____

2 Nam plays football for the under-8s team.

a Draw and write about something that you are **certain** would happen in the game.

b Draw and write about something that would be **impossible** to happen in the game.

3 Here are some things that Ling likes to do. Tick the things that are **likely to happen** in your class on a Tuesday.

read a book		eat lunch	
sit on a chair		play with water	
go to the library		have a different teacher	

4 What do you like to do at school, but which is **unlikely** to happen today? _____

_____

Unit
29 **Chance** (TRB pp. 134–137)
Chance MA1-18SP recognises and describes the element of chance in everyday events

119

A Bunch of Grapes

You will need: counters

Use counters to help you work out each problem.

1 There were 24 grapes in a bunch. Draw a quarter of the grapes on each plate.

$\dfrac{1}{4}$ of 24 is _____

2 There were 40 grapes in a bunch. Draw an eighth of the grapes on each plate.

$\dfrac{1}{8}$ of 40 is _____

3 There were 32 grapes in a bunch. Draw half of the grapes on each plate.

$\dfrac{1}{2}$ of 32 is _____

4 There were 16 grapes in a bunch. Draw a quarter of the grapes on each plate.

$\dfrac{1}{4}$ of 16 is _____

5 There were 32 grapes in a bunch. Draw plates, and show an eighth of the grapes on each plate.

$\dfrac{1}{8}$ of 32 is _____

6 There were 40 grapes in a bunch. Draw plates, and show a quarter of the grapes on each plate.

$\dfrac{1}{4}$ of 40 is _____

More About Fractions (TRB pp. 138–141)
Fractions and decimals MA1-7NA represents and models halves, quarters and eighths

Number Sentences and Pictures

1 Draw a picture to match each of the number sentences, and work out the answer.

$\frac{1}{8}$ of 16 = _____

$\frac{1}{2}$ of 14 = _____

$\frac{1}{2}$ of 18 = _____

$\frac{1}{4}$ of 8 = _____

2 Write a number sentence to match each picture, then work out the answer.

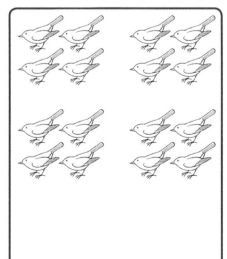

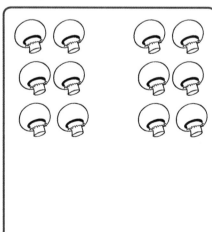

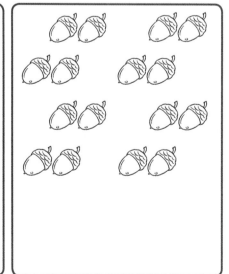

Unit 30 **More About Fractions** (TRB pp. 138–141)
Fractions and decimals MA1-7NA represents and models halves, quarters and eighths

121

Jelly Beans

You will need: coloured counters

1 Work out what a packet of jelly beans might look like if:

- one eighth of the jelly beans were yellow
- one quarter of the jelly beans were red.

Use some counters to help you.

Draw what some packets of jelly beans might look like.

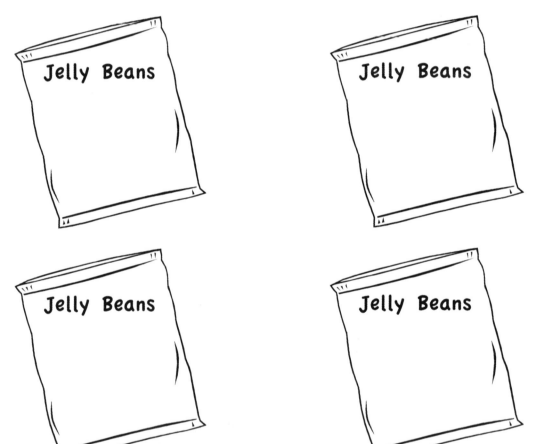

2 Explain how you know that you have drawn the correct number of yellow and red jelly beans in each packet in Question 1.

_____

_____

_____

STUDENT ASSESSMENT

1 a Jane had 16 marbles.

She gave $\frac{1}{4}$ to Jay, $\frac{1}{2}$ to Jim and $\frac{1}{8}$ to Jen.

Show how many each child received.

Jay   Jim   Jen

b Write the number sentence for each bag of marbles.

Jay: _____

Jim: _____

Jen: _____

2 Jan had a bag of marbles. One quarter of the marbles were red. Draw 3 ways her bag of marbles may have looked.

Unit
30
More About Fractions (TRB pp. 138–141)
Fractions and decimals MA1-7NA represents and models halves, quarters and eighths

123

Maths Glossary

Numbers

0	zero	**10**	ten	**20**	twenty
1	one	**11**	eleven	**30**	thirty
2	two	**12**	twelve	**40**	forty
3	three	**13**	thirteen	**50**	fifty
4	four	**14**	fourteen	**60**	sixty
5	five	**15**	fifteen	**70**	seventy
6	six	**16**	sixteen	**80**	eighty
7	seven	**17**	seventeen	**90**	ninety
8	eight	**18**	eighteen	**100**	one hundred
9	nine	**19**	nineteen	**1000**	one thousand

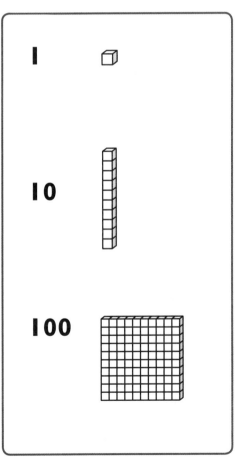

1

10

100

Mathematical Terms and Signs

+ addition
(and, altogether, combined with, plus)

− subtraction
(take away, difference)

✗ multiplication
(groups of)

÷ division
(shared between, how many groups of?)

= equals

Maths Glossary

Fractions

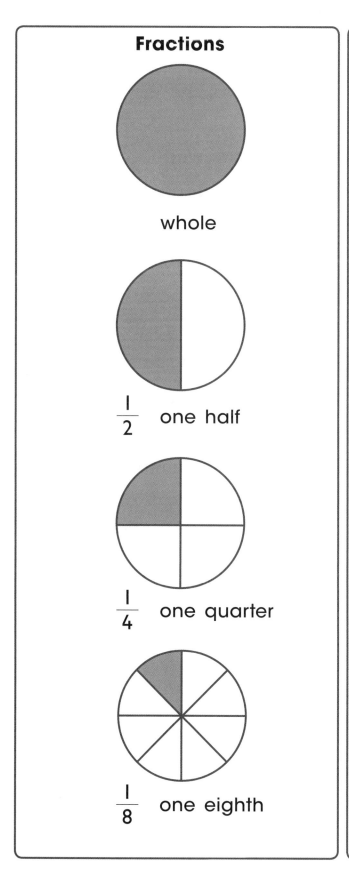

whole

$\dfrac{1}{2}$ one half

$\dfrac{1}{4}$ one quarter

$\dfrac{1}{8}$ one eighth

Money

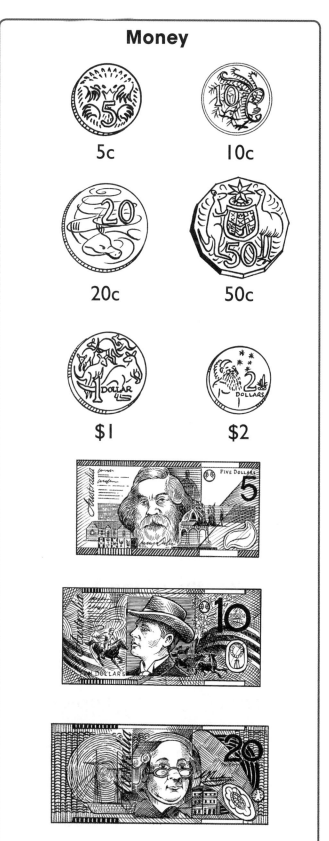

5c

10c

20c

50c

$1

$2

Maths Glossary

Measurement

Time

3 o'clock   half-past 9

quarter-past 10   quarter-to 5

Days of the Week

Monday

Tuesday

Wednesday

Thursday

Friday

Saturday

Sunday

Seasons

Spring

Summer

Autumn

Winter

Months of the Year

January	July
February	August
March	September
April	October
May	November
June	December

Length

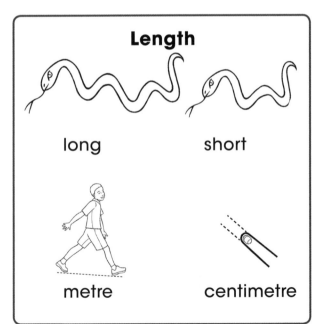

long   short

metre   centimetre

Maths Glossary

Measurement

Mass

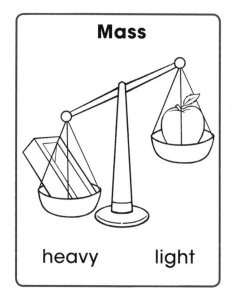

heavy   light

Capacity

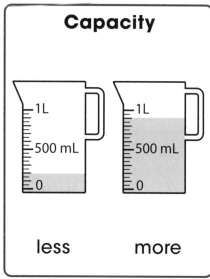

less   more

Area

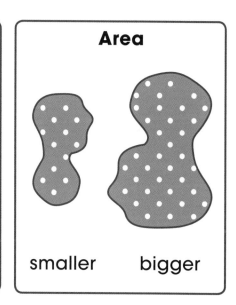

smaller   bigger

2D Shapes

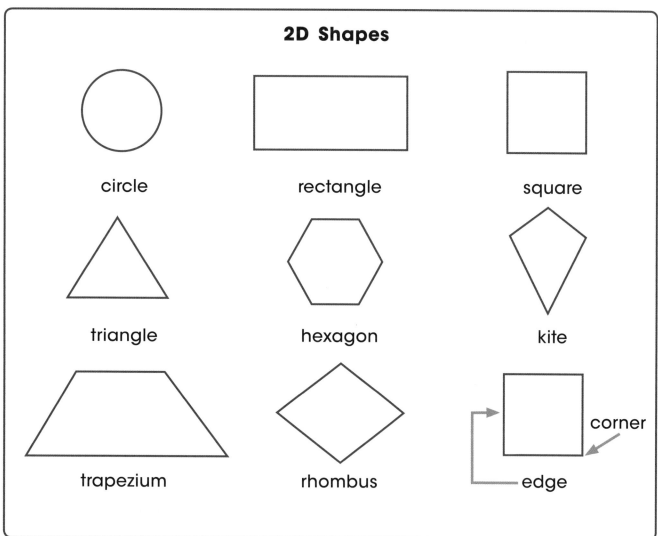

circle   rectangle   square

triangle   hexagon   kite

trapezium   rhombus   corner   edge

Maths Glossary

Measurement

Transformation

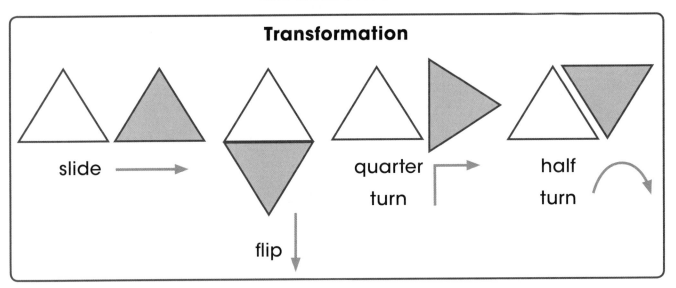

slide ⟶

flip

quarter turn

half turn

3D Objects

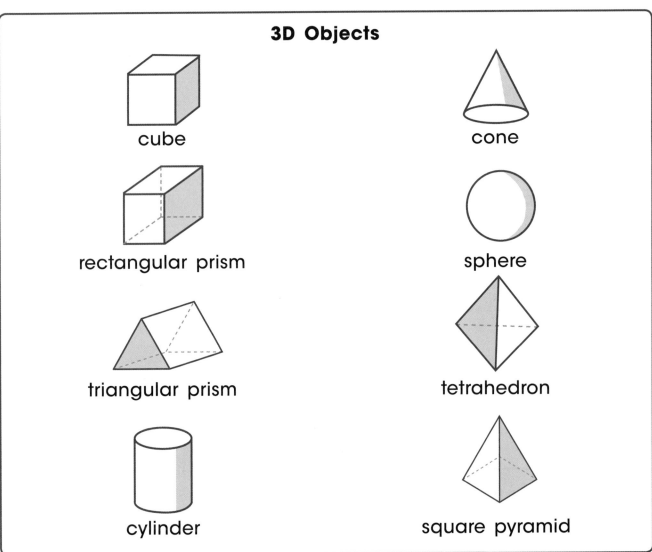

cube

cone

rectangular prism

sphere

triangular prism

tetrahedron

cylinder

square pyramid